Genetic Elements in *Escherichia coli*

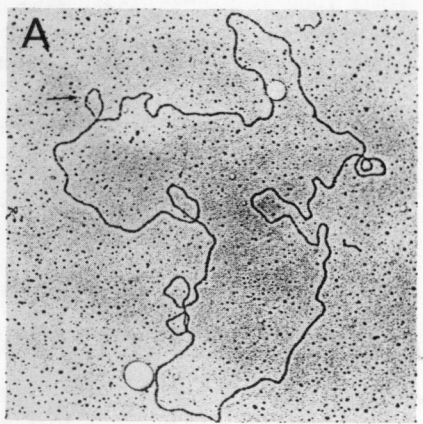

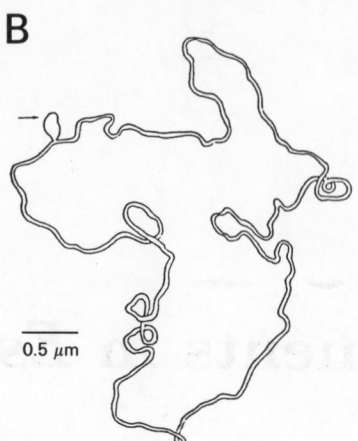

0.5 μm

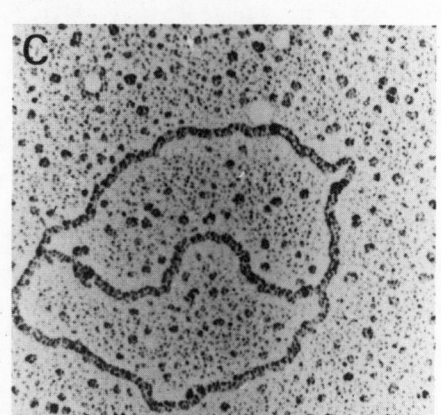

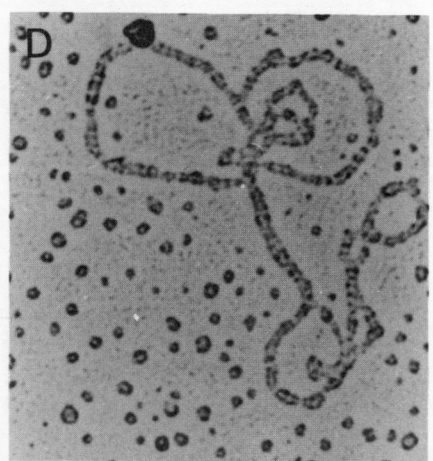

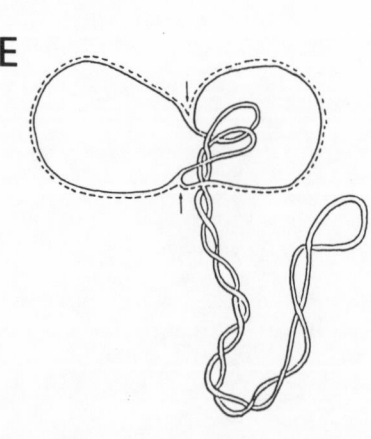

Frontispiece

UPPER: A heteroduplex molecule
Micrograph (A) and tracing (B) of a heteroduplex between plasmids RP4 and R68.45. These conjugal plasmids, isolated from *Pseudomonas aeruginosa,* carry genes conferring resistance to ampicillin, Kanamycin and tetracycline and have contour lengths of 19.1 μm and 19.7 μm respectively.

Note that the heteroduplex molecule is completely double stranded except for a single-stranded loop 0.6 μm long (arrowed). Thus R68.45 is probably an RP4 plasmid with a 0.6 μm ($\approx$ 1.8 kb) insertion mutation.

Electron micrograph kindly provided by Dr Hans-Joachim Burkardt.

LOWER: Replicating molecules of ColE1 DNA
ColE1 DNA replicates unidirectionally in both the open circular and covalently closed (supercoiled) forms. The partially replicated intermediates produced by replication of the open circular form are typical θ-type molecules (C) but when the covalently closed form replicates only the newly replicated DNA is untwisted and the unreplicated part of the molecule remains supercoiled (D).

At the right (E) is an interpretation of the 'butterfly' intermediate shown in C. The newly synthesised strand of DNA is represented by the dotted line; one of the forks (arrowed) is the replication origin.

Micrographs by the courtesy of Dr Ronald R. Helinski.

MACMILLAN MOLECULAR BIOLOGY SERIES
Series Editor: Dr C. J. Skidmore

Genetic Elements in *Escherichia coli*

P. F. Smith-Keary
Fellow and Senior Lecturer in Genetics, Trinity College, Dublin

MACMILLAN
EDUCATION

First published 1988

Published by
MACMILLAN EDUCATION LTD
Houndmills, Basingstoke, Hampshire RG21 2XS
and London
Companies and representatives
throughout the world

Printed in Hong Kong

British Library Cataloguing in Publication Data
Smith-Keary, P. F.
    Genetics of Escherichia Coli.—
    (Macmillan Molecular biology series).
    1. Escherichia coli—Genetics
    I. Title
    589.9′5   QH470.E8

    ISBN 0–333–44267–9
    ISBN 0–333–44268–7 Pbk

To Jean

By the same author
Genetic Structure and Function

CONTENTS

SERIES EDITOR'S PREFACE

The aim of this Series is to provide authoritative texts of a manageable size suitable for advanced undergraduate and postgraduate courses. Each volume will interpret a defined area of biology, as might be dealt with as a course unit, in the light of molecular research.

The growth of molecular biology in the ten years since the advent of recombinant DNA techniques has left few areas of biology unaffected. The information explosion that this has caused has made it difficult for large texts to keep up with the latest advances while retaining a proper treatment of the basics.

These books are thus a timely contribution to the resources of the student of biology. The molecular details are presented, clearly and concisely, in the context of the biological system. For, while there is no biology without molecules, there is more to biology than molecular biology. We do not intend to reduce biological phenomena to no more than molecular phenomena, but to point towards the synthesis of the biological and the molecular which marks the way forward in the life sciences.

C. J. Skidmore

PREFACE

In recent years molecular biology has acquired new dimensions largely as a result of the outstanding advances in molecular genetics, particularly the development of methods of analysis at the molecular level, culminating in our ability to isolate almost any gene from any organism, to purify it, sequence it, mutate it at will and then re-introduce it into another cell (often the bacterium *Escherichia coli*) where it can be expressed and produce protein. This book is intended to familiarise students of molecular biology with some of the fundamentals of bacterial and phage genetics and so to build a foundation for more advanced molecular genetic studies such as gene cloning, the use of restriction enzymes and recombinant DNA technology. As indicated by the title, I have (very largely) confined myself to considering *Escherichia coli* and its phages; *E. coli* is still the most widely used bacterium in molecular genetic studies and most of the concepts applied to *E. coli* apply equally to other bacterial species, particularly to the closely related species *Salmonella typhimurium*. Once, and only once, I have strayed into the field of fungal genetics as some of the most revealing information on the molecular mechanism of recombination — something fundamental to all genetic studies — has come from studies on gene conversion in the Ascomycetes.

I have assumed that the reader has taken introductory courses in biology and in genetics and is familiar with the basic principles of genetic structure and function in eucaryotes, including the construction of simple linkage maps in a diploid organism; the reader who has forgotten any of the jargon of genetics will find explanations of most of the technical terms in the Glossary.

In a book of limited size, the choice of topics is never simple and I have necessarily restricted myself to the particular aspects of bacterial and phage genetics that are likely to be the most useful to the intending molecular geneticist or molecular biologist; consequently, I have described the structure and mechanistic properties (replication, recombination, transposition and transmission) of the most important genetic elements of *E. coli* but have almost completely omitted the important fields of gene expression, recombinant DNA technology and the applications of molecular genetics in the present day society. For the same reasons, some topics are considered in more depth than others. For example, insertion sequences and transposons are described in some detail as they are of considerable importance to molecular biologists while the mutational process is only briefly considered; although mutational theory is of great interest and importance, the principal use of mutation to the molecular biologist is as a tool to produce particular mutant phenotypes.

I have tried to arrange the material in a reasonably logical sequence, explaining new terms and developing new concepts as they arise. However, many readers will follow their own sequence of reading and to assist them the text is extensively cross-referenced and most of the terms and concepts are also explained in the Glossary. Some further material, which though relevant lies outside the mainstream of the text, is presented in the form of 'boxes'.

Instead of summaries at the end of each chapter, there is a set of self-assessing questions and problems; for the most part these relate to the most important material presented in that chapter, but a few of the questions require an understanding of previous chapters and these serve to link together closely related aspects of microbial and molecular genetics.

Since this book is primarily intended for molecular biologists rather than for geneticists, very few original references are cited. Instead, at the end of each chapter, there are selected references to review-type articles and monographs; these not only provide more advanced, but comparatively easy-to-understand, further reading but also cite many of the most important original research papers referred to in the text. Some further suggestions for more general reading are given at the end of the book. I firmly believe that a student can only appreciate the complexities of present day molecular genetics if he is familiar with some of the key experiments from past years. These were sometimes very simple but always very elegant and intellectually satisfying, and they not only illustrate the development of new analytical methods but, more importantly, the conclusions made from them have led to the development of what we now regard as established concepts. Space has prevented any detailed consideration of this important earlier work but many of the original key papers are included in the collections of reprints also listed at the end of the book.

I hope the reader will get as much pleasure and benefit from reading this book as I have from writing it.

P. F. Smith-Keary

ACKNOWLEDGEMENTS

This book would not have been written but for the encouragement given to me by the Members of the Department of Genetics and I am most grateful for the advice and criticism they have given during its writing; in particular I would thank Dr Paul Sharp for his valuable assistance in correcting the proofs.

I extend my most sincere thanks to Dr H.-J. Burkhardt and Dr D. R. Helinski for providing the electron micrographs reproduced in the frontispiece.

ABBREVIATIONS AND SYMBOLS

(Note: *E. coli* gene symbols are explained in table 4.3; other gene symbols are explained in the text.)

A	adenosine
AP	apurinic or apyrimidinic site
2-AP or AP	2-aminopurine
Asp	aspartic acid
bp	base pair
5-BU or BU	5-bromouracil
C	cytidine
Col	a colicin plasmid
cos	cohesive and sequence of lambda
Δ	deletion
DNA	deoxyribonucleic acid
DNase	deoxyribonuclease
DP	DNA polymerase
F	the *E. coli* fertility plasmid
G	guanosine
Gly	glycine
HA	hydroxylamine
Hfr	high-frequency recombination
IR	inverted repeat sequence
IS	insertion sequence
kb	kilobases or kilobase pairs
Lac	lactose
Leu	leucine

μm	micrometre (10^{-6} metres)
Met	methionine
mRNA	messenger RNA
nm	nanometre (10^{-9} metres)
O	operator region
ori	replication origin
P	promoter region
PRR	post-replicational repair
R	a resistance plasmid
RNA	ribonucleic acid
RNase	ribonuclease
RTF	resistance transfer factor
SSR	site-specific recombination
T	thymidine
Tn	transposon
tra	transfer operon
tRNA	transfer RNA
Trp	tryptophan
Tyr	tyrosine
U	uridine
UV	ultraviolet irradiation

STRUCTURE AND REPLICATION OF GENETIC ELEMENTS IN *E. COLI*

1.1 Introduction

Escherichia coli, the common gut bacterium, occurs naturally in both man and animals as part of the bacterial flora of the intestine and although normally harmless, exceptional strains are pathogenic and cause diarrhoea in travellers, infants and livestock. It has been used in genetic studies for many years and as early as 1907 **R. P. Mansini** isolated a lactose non-fermenting (lac^-) mutant which produced lactose-fermenting (lac^+) derivatives at a very high frequency and, since then, particularly during the last 40 years, its genetic structure and function have been studied in the greatest detail. Many important problems of molecular genetics have been solved directly as a result of studies using *E. coli* or its phages (table 1.1) and it is now extensively used as a host for cloning-vectors.

A closely related gut bacterium, *Salmonella typhimurium*, has been extensively used in parallel studies and its genetic organisation is very similar to that of *E. coli*. In nature, strains of *S. typhimurium* are the most important single cause of food poisoning in man and animals, but the strains used in genetic experiments are usually quite harmless.

1.2 The structure of *E. coli*

E. coli is a procaryote and, unlike eucaryotes or higher organisms, it does not have a nuclear membrane, there are neither chloroplasts nor mitochondria and a spindle does not form at cell division. It is a rod

Table 1.1
Some notable discoveries made using *E. coli* and its phages

1907	R. MANSINI isolates highly mutable Lac⁻ mutants of *E. coli*
1943	SALVADOR LURIA & MAX DELBRUCK disprove the theory of acquired immunity and show that changes to phage resistance are the consequence of spontaneous mutations
1946	JOSHUA LEDERBERG & EDWARD TATUM discover recombination in *E. coli*
1953	WILLIAM HAYES shows that conjugation in *E. coli* is mediated by a transmissible factor, F
1956	ELIE WOLLMAN, FRANCOIS JACOB & WILLIAM HAYES establish the nature of the Hfr × F⁻ mating process and devise novel methods for mapping the bacterial chromosome
1956	ARTHUR KORNBERG pioneers the *in vitro* (enzymatic) synthesis of DNA in a system using extracts from *E. coli*
1957	MEHRAN GOULIAN, ARTHUR KORNBERG & ROBERT SINSHEIMER synthesise *in vitro* the first biologically active DNA (using ϕX174)
1958	MATTHEW MESELSON & FRANKLIN STAHL demonstrate the semi-conservative replication of the *E. coli* chromosome
1961	FRANCIS CRICK & colleagues establish the general nature of the genetic code using the *rII* system of phage T4
1961	SYDNEY BRENNER, FRANCOIS JACOB & MATTHEW MESELSON confirm the messenger RNA hypothesis using T2 infected cells of *E. coli*
1961	FRANCOIS JACOB & JACQUES MONOD propose the operon model for the genetic control of protein synthesis
1963	JOHN CAIRNS visualises replicating *E. coli* chromosomes by autoradiography
1966	GOBIND KHORANA achieves the first *chemical* synthesis of a gene
1967	BORIS SHAPIRO, PETER STARLINGER, HEINZ SAEDLER, ELKE JORDAN and MICHAEL MALAMY discover insertion sequences
1974	R. W. HEDGES & A. E. JACOB discover transposons
1977	K. ITAKURA and colleagues use an *E. coli* host and a chimeric plasmid constructed by recombinant DNA technology to clone a human gene, and produce the first synthetic human protein, the hormone somatostatin
1978	DAVID GOEDDEL and coworkers clone the human insulin gene in *E. coli* and produce insulin

about 2 μm long and 0.7 μm in diameter and so is very much smaller than the average cell of a eucaryote.

The outermost layer of the *E. coli* cell is the **cell envelope**; this corresponds to the cell wall of a plant cell and in turn it consists of an outer membrane, a peptidoglycan layer which adds rigidity to the envelope, a periplasmic region and an inner membrane. Inside the cell envelope is the cytoplasm, filled with free ribosomes, molecules of transfer RNA, a wide variety of enzymes and many different metabolic products; embedded in this RNA-rich surround is a central compact region, the **nucleoid**, in which is located the chromosomal DNA.

Three types of appendage may be found attached to the outer membrane:

(i) **Flagellae** These are responsible for motility of the cell and there are between 0 and 100 per cell, each about 20 μm long and 20 nm in diameter.

(ii) **Common pili** or fimbriae. These are smaller than the flagellae (about 1 μm × 10 nm) and can be very numerous.

(iii) **The F- or sex-pilus.** A single sex-pilus (about 1–2 μm × 10 nm) may be present on cells that harbour an F factor or other conjugative plasmid (sections 1.4 and 3.2). These pili are encoded by genes present on the plasmid and are essential to the conjugation process and the transfer of genetic information (section 4.2). They also have the unusual role of providing the specific receptor sites for certain bacteriophages such as MS2.

1.3 The chromosome of *E. coli*

By comparison with a eucaryotic chromosome, the chromosome of *E. coli* is a very simple structure consisting of a single covalently bonded circular molecule of DNA with a contour length of 1100 μm (about 3800 kb of DNA). Although this is very much less than the amount of DNA in a typical eucaryotic cell, the molecule is still very long in relation to the size of the cell and it must be very tightly twisted in order to be accommodated within the cell. We have little idea as to how a DNA molecule can function when it is so tightly packaged, particularly since both replication and transcription involve separation of the two tightly coiled strands of the DNA double helix. Prior to 1976 it was thought that the *E. coli* chromosome was simply a naked molecule of DNA but we now know that there are associated proteins which complex with the DNA and maintain it in a highly condensed state. In 1976 **Jack Griffith** successfully isolated intact nucleoids from gently disrupted cells of *E. coli* and visualised these by electron microscopy. His studies revealed, firstly, that the chromosomal DNA was condensed into a bumpy fibre with a diameter of about 12 nm (the diameter of a DNA molecule is 2 nm) and, secondly, that this fibre was folded along its circumference into about 100 loops (figure 1.1). Each loop contained about 40 kb of DNA and had a contour length of 2 to 2.5 μm; since 40 kb of fully extended DNA measures 13.6 μm, the DNA within the fibre must be condensed about 6.5 fold. Where it had been slightly stretched, the bumpy fibre resolved into a series of beads connected by a very fine filament. The organisation of the DNA within the beads is

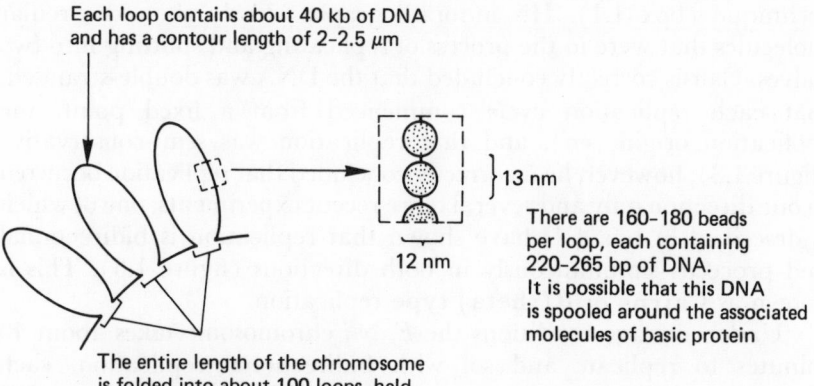

Each loop contains about 40 kb of DNA and has a contour length of 2–2.5 μm

13 nm

12 nm

There are 160–180 beads per loop, each containing 220–265 bp of DNA. It is possible that this DNA is spooled around the associated molecules of basic protein

The entire length of the chromosome is folded into about 100 loops, held together at their bases in some unknown way

Figure 1.1
The nucleoid of *E. coli*

Figure 1.2
The chromosome of *E. coli* is attached to the cell membrane

In a non-dividing cell (1) the chromosome is attached to an invagination of the cell membrane, the mesosome. After the chromosome has replicated (2) the membrane grows between the two points at which the daughter chromosomes are attached (3), ensuring that one goes into each daughter cell. After the completion of cell division (4) each daughter cell has a single chromosome attached to its own membrane.

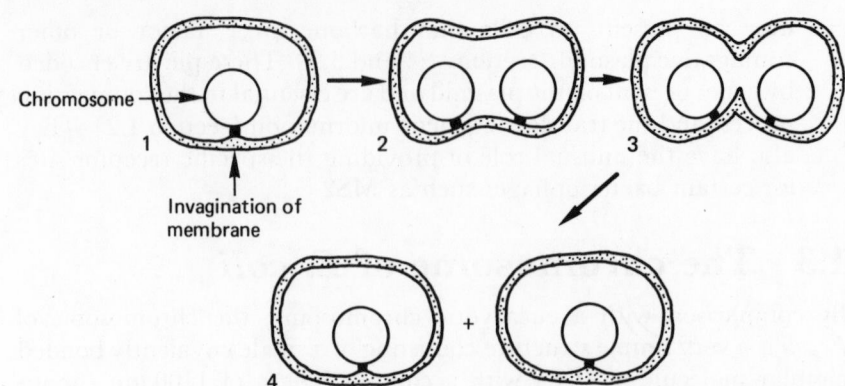

not known but the beads have a striking similarity to the nucleosomes found in eucaryotic chromosomes. Four specific nucleoid-associated proteins have been identified and it is likely that these maintain this structure in the same way as specific histone and non-histone proteins are responsible for the structural organisation of the eucaryotic chromosome.

Electron micrographs have shown that the chromosomal DNA is attached to an invaginated part of the cell membrane known as the **mesosome** and it is probable that this membrane plays a crucial role in ensuring the regular segregation of the two daughter chromosomes whenever the cell divides (figure 1.2). Thus the mesosome achieves the same purpose as the mitotic spindle in eucaryotes.

1.3.1 Replication of the *E. coli* chromosome

Initially the evidence from electron microscopy suggested that all DNA molecules were linear and had two free ends but, as the methods for extracting DNA were improved, it became more likely that some molecules of DNA, including the chromosome of *E. coli*, were circular. The first definitive evidence that the *E. coli* chromosome was circular was obtained by **John Cairns** in 1963 using an autoradiographic technique (box 1.1). His autoradiographs clearly showed circular molecules that were in the process of replicating and splitting into two halves. Cairns correctly concluded that the DNA was double-stranded, that each replication cycle commenced from a fixed point, the replication origin (*ori*), and that replication was semi-conservative (figure 1.3); however, he incorrectly concluded that replication occurred in one direction only and several more recent experiments, one of which is described in box 1.1, have shown that replication is bidirectional and proceeds simultaneously in both directions (figure 1.3). This is known as **Cairns** or θ (**theta**) **type** replication.

Under average conditions the *E. coli* chromosome takes about 40 minutes to replicate and so, with bidirectional replication, each replication fork must move at the astonishing rate of about 800 nucleotides per second.

Box 1.1 Autoradiography of the *E. coli* chromosome

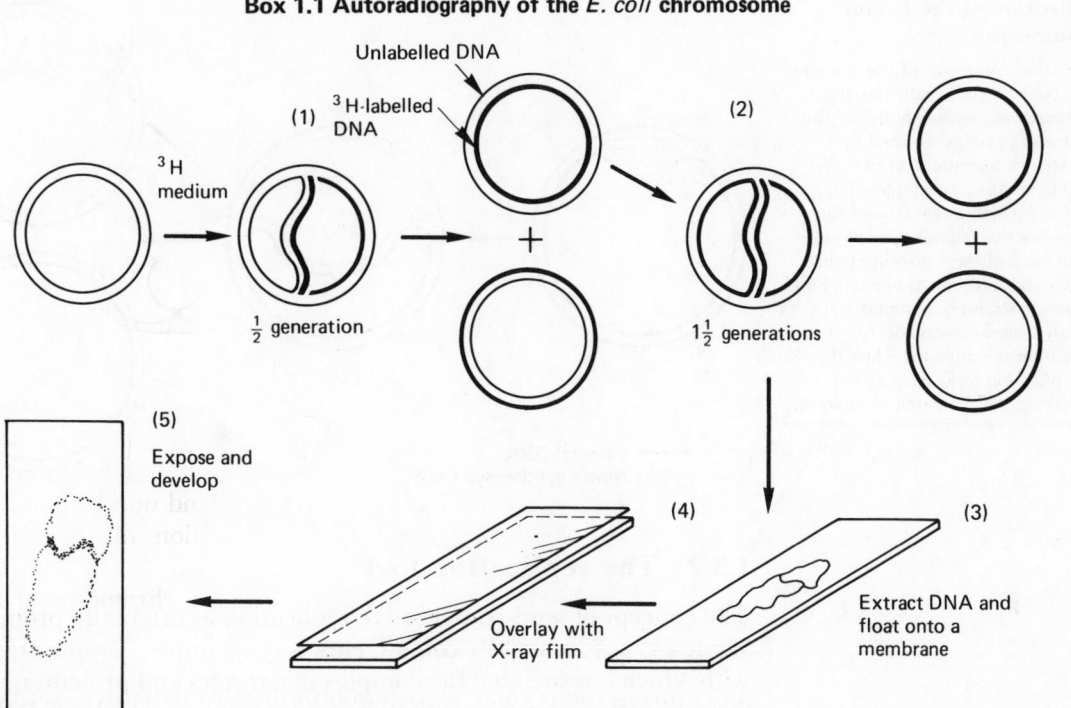

In 1963 John Cairns made the first autoradiographs of the *E. coli* chromosome. He transferred growing *E. coli* cells to a medium containing tritiated thymidine so that any newly synthesised DNA was ^{3}H-labelled (1). After 1–2 generations (2) the DNA was gently extracted, floated onto dialysis membranes, placed on a microscope slide (3), overlaid with a photographic emulsion (4) and exposed for 1–2 months. During this period the ^{3}H particles decay and emit β-particles; these expose the emulsion and, after development (5), each emission is recorded as a black spot. Since β-particles only travel about 1 μm, the position of the spots in the emulsion indicates the actual location of the tritium particles in the underlying chromosome. The resulting autoradiographs revealed circular molecules of DNA in the process of replicating while the density of the spots indicated whether one or both strands of DNA was ^{3}H-labelled. In this diagram the chromosome is shown as replicating bidirectionally (5) and the label is assumed to have been introduced exactly at the start of a replication cycle.

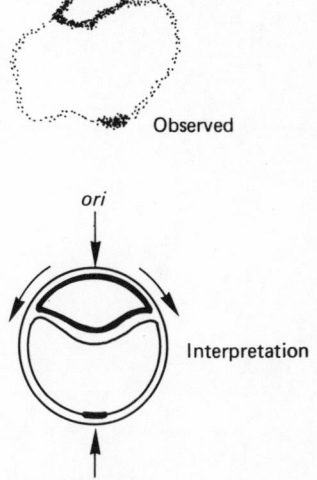

In 1973 Ray Rodriguez refined the method and was able to prove that the chromosome replicated bidirectionally. He labelled synchronously growing *E. coli* with a low activity tritium label, introduced just before the commencement of a replication cycle. Just before the end of this cycle he transferred the cells to a high activity ^{3}H medium and then, after a short interval, examined the chromosomes by autoradiography. As a result, the high activity tritium labelled the terminus of the first replication and the replication forks of the second replication cycle. The autoradiographs showed that both limbs of the theta structure are labelled, showing that there are two replication forks; furthermore, the terminus and the replication origin are at opposite sides of the circular molecule. These results are only expected if replication proceeds bidirectionally from a fixed origin.

Figure 1.3
**Replication of the *E. coli*
chromosome**

Replication commences from a unique
origin (*ori*) by the separation of the
two strands of the double helix. The
complex of proteins required for
replication is assembled at each Y-
junction, creating two replication
forks (←) which move around the
chromosome in opposite directions.
Within each of these growing points
the parental molecule is unwound and
each single strand of parental DNA is
replicated semi-conservatively.
Replication is completed when the
two replication forks meet at the
opposite side of the parental molecule.

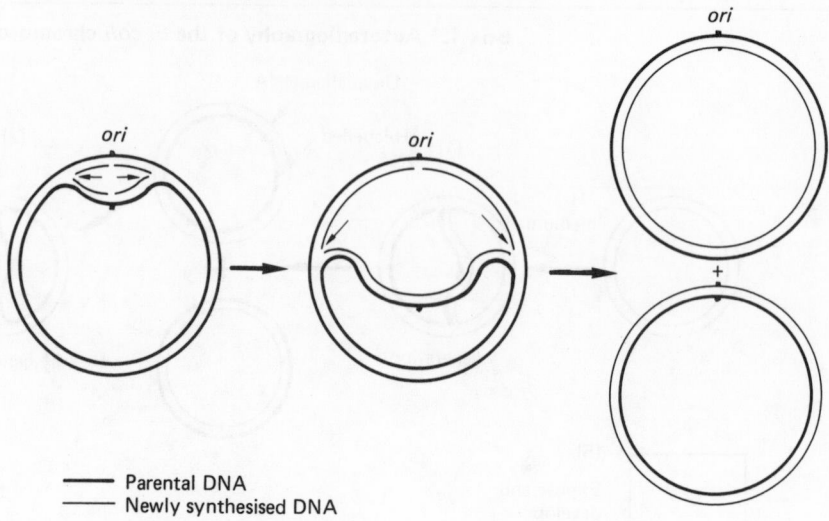

—— Parental DNA
—— Newly synthesised DNA

1.3.2   The replication fork

The concept of semi-conservative replication as originally proposed by
James Watson and Francis Crick is very simple. A replication fork,
with which is associated the complex of enzymes and proteins necessary
for replication, moves progressively along the molecule of DNA and
unwinds the two strands of the parental (template) molecule; at the
same time the newly synthesised and base complementary daughter
strands are elongated by successive nucleotides polymerising on to their
growing ends, each selected nucleotide being the complement of one
present on the parental strand. In practice, however, the replication
of DNA is one of the most complex biological processes and in *E. coli*
involves the protein products of at least 15 different genes (table 1.2).

The most important enzyme involved is **DNA-dependent DNA
polymerase III** (DNA polymerase III or DP III), the actual
replication enzyme, or replicase, which acts within the replication fork
to copy the information on the strands of parental DNA. However,
DP III not only requires a template strand to copy but it can **only**
polymerise deoxyribonucleotides on to an **existing** polynucleotide
chain (either DNA or RNA) and so, by itself, it is unable to initiate
replication. As a result, each independent replicated segment of DNA
requires a short strand of RNA, the **primer**, which provides the
polynucleotide sequence on to which DP III can then add deoxyribonucleotides. In *E. coli* these RNA primers are usually about five
nucleotides long and are normally synthesised by the enzyme **DNA
primase** which is able to polymerise ribonucleotides against a single-
stranded DNA template. Another enzyme sometimes involved in
priming is **DNA-dependent RNA polymerase**; this is the enzyme
that transcribes RNA against a single-stranded DNA template and it
is known to prime the replication of the ColE1 plasmid and of the
single-stranded DNA phage M13.

Enzyme and protein	Gene(s)	Properties
1. DNA polymerases		1. All polymerise deoxyribo-nucleotides on to the 3′–OH end of an existing oligonucleotide, using the complementary DNA strand as a template
		2. All have 3′→5′ exonuclease activity and can digest either strand of duplex DNA from an exposed 3′–OH end
(a) DP I	*polA*	3a. Also has 5′→3′ exonuclease activity and can digest either strand of duplex DNA from an exposed 5′ end
(b) DP II	*polB*	3b. Probably involved in DNA repair
(c) DP III	*polC* (= *dnaE*), *dnaX*, *dnaZ*, and others	3c. Also has 5′→3′ exonuclease activity but this will act only on single-stranded DNA
2. DNA primase	*dnaG*	1. Polymerises ribonucleotides against a single-stranded DNA template
		2. Primes the synthesis of Okazaki fragments and normally primes the initiation of DNA replication
3. RNA polymerase	*rpoA*, *rpoB*, *rpoC*, *rpoD*.	The normal transcription enzyme, but it may also prime the initiation of DNA replication
4. DNA ligase	*lig*	Forms phosphodiester bonds between two adjacent and unlinked nucleotides in a DNA duplex
5. DNA gyrase	*gyrA*, *gyrB*	Removes negative supercoils from circular molecules of DNA (that is, unwinds DNA)
6. Rep protein	*rep*	Assists in unwinding the DNA duplex. Not essential for *E. coli* replication but is necessary for the replication of some coliphages
7. Single-strand binding protein	*ssb*	Binds to and protects the single-stranded DNA exposed at the replication fork

Table 1.2
Properties of some of the enzymes and other proteins involved in the replication of the DNA of *E. coli*

Furthermore, DP III can only polymerise deoxyribonucleotides on to the 3′–OH end of a DNA strand and so can **only** operate in the 5′→3′ direction; thus while DP III could replicate one strand of the duplex strictly according to the concepts of Watson and Crick, it could not replicate the other strand in the same way as this would involve synthesis in the 3′→5′ direction. The answer to this dilemma is that this strand, the so-called **lagging** strand, is replicated **discontinuously**. Short pieces of DNA, about 1000 nucleotides long, are individually primed and templated in the 5′→3′ direction and then enzymatically joined on to the growing chain (figure 1.4); although the overall extension of this strand is in the 3′→5′ direction, the actual synthesis is in the $5' \rightarrow 3'$ direction. These short discontinuous synthesised segments of DNA are known as **Okazaki fragments**, after **Reiji Okazaki** who first detected them in 1968. Finally, the individually completed segments of DNA are joined together by **DNA ligase**, which can join the $3'$ end of an Okazaki fragment on to the $5'$ end of the preceding fragment by the formation of a $3':5'$-phosphodiester bond, so completing the continuity of the sugar–phosphate backbone of the newly synthesised strand.

However, before these single-strand gaps can be sealed by DNA ligase, any RNA primer sequences must be removed and replaced by the corresponding deoxyribonucleotides. This is achieved by **DNA polymerase I** (DP I), first isolated by **Arthur Kornberg** in 1958 and at that time thought, albeit incorrectly, to be the true replicase.

Figure 1.4
Priming and the discontinuous replication of DNA

The two strands of the double helix of DNA have partly separated and each is acting as a template for a new daughter strand. The replication fork is moving from right to left and its principal components are shown in the box at the top of the figure. The upper parental strand acts as a template for the leading strand, synthesised continuously in the 5' to 3' direction. The other daughter strand, the lagging strand, is synthesised in the form of short fragments which are then ligated together.

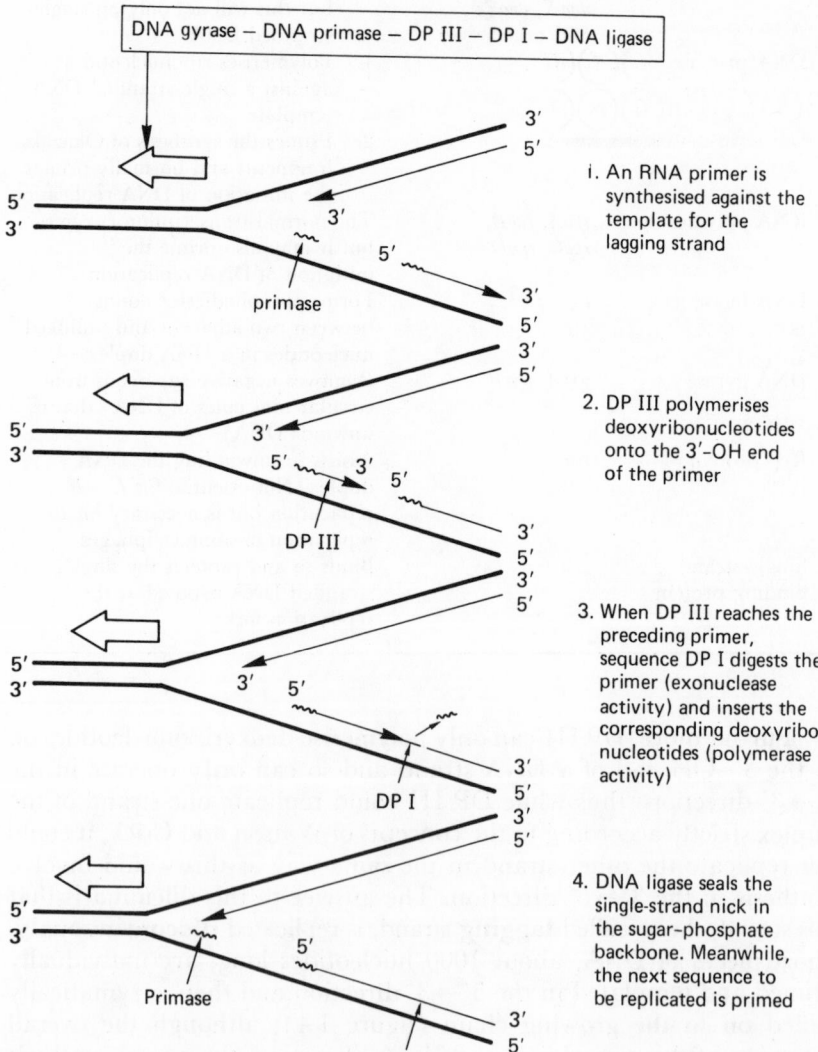

DNA gyrase – DNA primase – DP III – DP I – DNA ligase

i. An RNA primer is synthesised against the template for the lagging strand

primase

2. DP III polymerises deoxyribonucleotides onto the 3'-OH end of the primer

DP III

3. When DP III reaches the preceding primer, sequence DP I digests the primer (exonuclease activity) and inserts the corresponding deoxyribonucleotides (polymerase activity)

DP I

4. DNA ligase seals the single-strand nick in the sugar–phosphate backbone. Meanwhile, the next sequence to be replicated is primed

Primase

DNA ligase

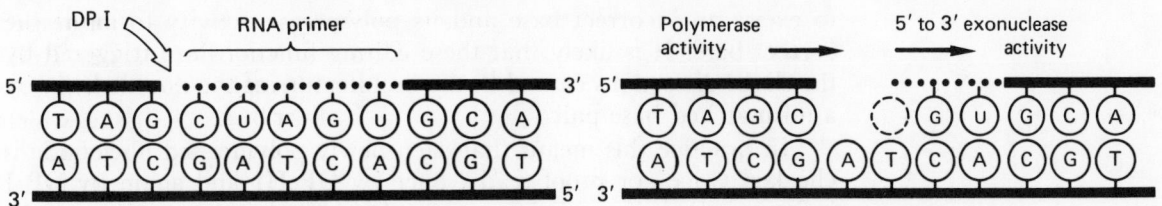

1. Primase has synthesised a 5-base-long RNA primer (CUAGU) to which is covalently bonded the newly synthesised Okazaki fragment (GCA . .). DP I recognises the free 5′ end of the RNA primer

2. The 5′ to 3′ exonuclease activity of DP I digests away the RNA primer. Simultaneously the polymerase activity of DP I polymerises the corresponding deoxyribonucleotides (CTAGT) on to the 3′ end of the preceding Okazaki fragment

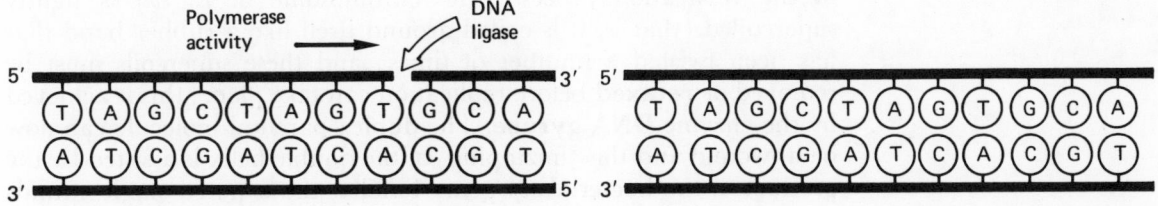

3. The primer has been removed and replaced by deoxyribonucleotides

4. The remaining single-strand gap has been sealed by DNA ligase

This enzyme can also polymerise nucleotides on to the 3′–OH terminus of an existing strand, although very much more slowly then DP III, and in addition has both 3′→5′ and 5′→3′ exonuclease activity. An **exonuclease** is an enzyme that can digest DNA commencing at the **free** end of a strand and, because DP I has both 3′ and 5′ exonuclease activity, it can progressively digest nucleotides from either an exposed 3′ or 5′ end. It is thought that DP I recognises the free 5′ ends of the primer and, using its 5′→3′ exonuclease activity, digests away the primer sequence; at the same time it uses its polymerase activity to fill in the resulting gaps with the corresponding deoxyribonucleotides (figure 1.5).

Clearly, extreme accuracy is essential if the genetic information is to be transmitted unchanged from one generation to the next and although DP III is extremely accurate in selecting the correct complementary nucleotides, it does occasionally make a mistake. However, DP III, like DP I, has 3′→5′ exonuclease activity and, whenever a mismatched nucleotide is incorporated, DP III can reverse direction, retreat in a direction opposite to that of synthesis, use this exonuclease activity to cut out the incorrect nucleotide and then continue synthesis.

If by any chance the incorrect base is not immediately corrected in this way by DP III, it can still be corrected by DP I. As DP III progresses along the replicating molecule it is very closely followed by a molecule of DP I and this also can use its 3′→5′ exonuclease activity

Figure 1.5
Primer removal and gap sealing

Note that these events only occur along the discontinuously synthesised lagging strand.

to excise an incorrect base and its polymerase activity to insert the correct base. It is likely that these editing functions are triggered by the slight distortions caused in the architecture of the double helix by a mismatched base pair.

In practice this means that each newly polymerised nucleotide is checked twice (or proof read), once by DP III and again by DP I before synthesis is continued, so greatly increasing the accuracy of the replication process.

The overall pattern of replication is described as **semi-discontinuous** because one daughter strand (the leading strand) is synthesised continuously and the other strand (the lagging strand) is synthesised discontinuously.

We may briefly note three other proteins that play important roles in the replication process. The chromosome of *E. coli* is tightly supercoiled (that is, it is coiled around itself like a rubber band that has been twisted a number of times) and these supercoils must be removed or **relaxed** before replication can take place; this is achieved by the enzyme **DNA gyrase**. The duplex parental molecule can now be unwound and this 'unzipping' of the double helix is assisted by the product of the bacterial *rep* gene. Finally, the exposed single strands are stabilised by the binding of a special protein which specifically binds to single-stranded DNA; this single-strand binding protein (*ssb*) both protects the exposed single strands from nuclease attack and prevents them from reforming double-stranded DNA and blocking replication.

The symbols used for representing the genes and phenotypes of *E. coli* are explained in box 1.2, while all the genes referred to in the text are listed in table 4.3.

1.4 Insertion sequences, transposons and plasmids

In addition to the DNA comprising the chromosome, *E. coli* may contain molecules of at least four other species of DNA. These are: (1) insertion sequences (section 7.2), (2) transposons (section 7.3), (3) plasmids (chapter 3), and (4) bacteriophage DNA (section 1.5 and chapter 8).

Insertion sequences (IS) and **transposons** form a remarkable class of genetic element able to **transpose** from one position in the genome to another, and although they are found principally on plasmids they also occur on bacterial and bacteriophage genomes. IS elements are specific DNA sequences usually 750–1500 bp long and the only proteins they encode are the specific **transposases** involved in the transposition process. They are able to insert into molecules of DNA more or less at random and they will inactivate any structural gene into which they transpose; furthermore, since they carry a variety of transcriptional start and stop signals (promoters and terminators) they can have a profound effect on the expression of any adjacent genes. Transposons are much larger elements, typically between 2.5 and 9 kb long; they also encode a variety of important characteristics such as

---

Box 1.2 Genetic nomenclature of *E. coli*

Genes in *E. coli* (and in most bacteria) are symbolised by a three-letter italicised abbreviation or acronym of the major character they control. Thus *lac* symbolises any one of the three structural genes involved in the utilisation of lactose as a carbon source, and *trp* represents any one of the genes concerned with the bio-synthesis of tryptophan. When more than one gene affects the same character, each is distinguished by adding a capital letter to the symbol; thus *lacZ*, *lacY* and *lacA* symbolise the three structural genes within the *lac* operon (see box 4.1). Finally, different mutants affecting the same character are distinguished by adding a number to the gene symbol; thus *lac-15* represents the 15th mutation isolated within the *lac* operon and implies that the particular gene within which the mutation occurred has not yet been established, while *lacZ1* and *trpD1081* identify particular mutations within the *lacZ* and *trpD* genes.

In this text a superscript ($^-$) is frequently added to the symbol for a mutant gene; this is not standard practice but it can help in avoiding confusion. Functional wild type genes are usually distinguished by adding a superscript ($^+$); thus a particular mutant *lacZ* gene can be represented *lacZ2* or *lacZ2$^-$* while the correspond-ing wild type gene is symbolised *lac$^+$* or *lacZ$^+$*. If a strain is designated as just *lacZ2$^-$* (for example) the impli-cation is that all other relevant but unspecified genes are wild type (*lacY$^+$* and *lacA$^+$* for example).

Note, however, that when there is no ambiguity it is customary to refer to the *lacZ* gene and the *lacZ* gene product and to omit the ($^+$) superscripts.

Phenotypes.   The phenotype, or observable characteristics of a strain, is represented by a three-letter non-italicised symbol. Thus Lac$^+$ indicates a strain able to ferment lactose and Trp$^-$ a strain that can only grow if tryptophan is present in the medium.

All the gene symbols used are explained in table 4.3.

---

resistance to antibiotics and the ability to produce certain toxins and, because they are a frequent component of certain transmissible plasmids, both they and the plasmids that carry them are of considerable importance in medical and veterinary practice. The IS elements and transposons only exist as components of DNA molecules which are able to replicate.

An important feature of insertion sequences and transposons is that the two ends of each element form a pair of **inverted repeat sequences** (IR sequences). This means that the sequence of nucleotide pairs at one end of the element is repeated at the other end but in the reverse order, thus:

$$5' \quad C\ C\ A\ T\ T \ldots\ldots A\ A\ T\ G\ G \quad 3'$$
$$3' \quad G\ G\ T\ A\ A \ldots\ldots T\ T\ A\ C\ C \quad 5'$$

These inverted repeats are highly specific nucleotide sequences and, since they are the sequences recognised by the transposases, they are responsible for many of the unusual properties of transposable elements.

The **plasmids** found in *E. coli* are covalently closed circular (CCC) molecules of DNA and, in contrast to transposable genetic elements, each is an independent unit of replication or **replicon**. They exist alongside the bacterial chromosome and, like it, are stably inherited with at least one copy of each plasmid being transmitted to each daughter cell. Some plasmids, such as F and R, are **conjugative** plasmids and are self-transmissible not only from one *E. coli* to another but also between several other species of gram-negative bacteria. These plasmids encode a protein tube, the **sex-pilus**, which can join together two bacterial cells one of which harbours the plasmid, and this, in turn, enables the formation of a cytoplasmic bridge through which the plasmid DNA can be transmitted from one cell to the other (section 3.2.3). The F plasmid is also able to integrate into the bacterial chromosome and it can then promote the transfer of segments of the bacterial chromosome from one strain of *E. coli* to another (section 4.2); plasmids like F, which can either exist autonomously or integrated into the chromosome of the host cell, are sometimes called **episomes**, although this term is no longer in common use. Other plasmids, such as ColE1, are **non-conjugative** although they can often be transferred from cell to cell if a conjugative plasmid is also present.

In general, the conjugative plasmids are large (37–120 kb) and there are only one or two copies per bacterium, while the non-conjugative plasmids are small (4.5–9 kb) and present in numerous copies.

Most plasmids are very stably inherited and less than one cell in 100 000 does not carry a copy of the plasmid. This suggests that there is some mechanism which ensures that each daughter cell receives at least one copy of the plasmid, at least in those instances when there is only one (or perhaps two) copies of the plasmid per cell. A very likely possibility is that each plasmid is attached to a specific site on the mesosome in the same way as the bacterial chromosome (figure 1.2); as the membrane grows during cell division so each pair of daughter sites becomes separated and one site from each pair, together with the attached chromosome or plasmid, passes into each daughter cell. This model, first suggested by **François Jacob, Sydney Brenner and François Cuzin** in 1963, has received renewed support in recent years.

1.4.1   Plasmid replication and the control of copy number

Although all plasmids, like the *E. coli* chromosome, replicate vegetatively by a θ-mode mechanism, the details of the process differ from one plasmid to another, particularly as to the way that replication is initiated and controlled.

Among the small non-conjugative plasmids the replication of ColE1 is the best understood and has been extensively studied by **Jun-ichi Tomizawa** and his colleagues; it uses the replication machinery of the host cell. Replication is initiated by DNA-dependent RNA polymerase (the enzyme normally involved in the transcription of DNA) transcribing the heavy (H) strand commencing at a promoter 555 bases upstream from the replication origin (figure 1.6) and proceeding past the replication origin, *oriV*; this RNA is known as **RNA II** and it now hybridises with the H-strand in the region of *oriV* (the two strands of the ColE1 duplex are referred to as the heavy and light strands since they can be separated by density gradient centrifugation—see box 2.1). Another enzyme, RNase H, now cuts this RNA within two bases of the replication origin thus removing the 3' tail of RNA II (figure 1.6(3)); the newly exposed 3' end of RNA II now acts as a primer

A. *The Initation of Replication*

Figure 1.6
The initiation of replication of ColE1

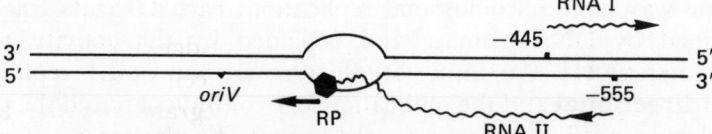

1. RNA polymerase transcribes the heavy (H) strand from a promoter 555 bp upstream from the replication origin. This RNA II transcript extends beyond *oriV*

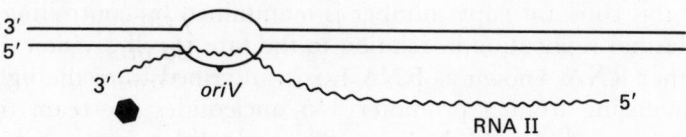

2. Transcription terminates and RNA II hybridises to the H-strand of DNA in the region of *oriV*

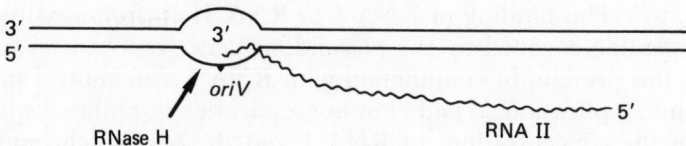

3. RNase H cuts off the tail of RNA II within two nucleotides of the replication origin

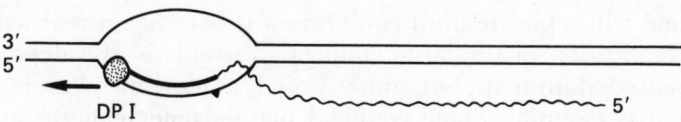

4. DP I adds deoxyribonucleotides on to the newly exposed end of RNA II

B. *The Control of Replicational Initiation*

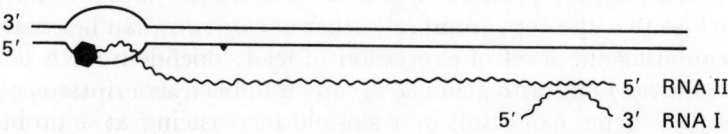

RNA I can bind to the 5′ end of RNA II by complementary base pairing. This may alter the structure of RNA II, prevent it from hybridising with the H-strand of DNA, and block the formation of the DNA–RNA substrate required for RNase H activity

for the addition of deoxyribonucleotides by DNA polymerase I, but after some 500 nucleotides have been added it seems that DNA polymerase III takes over and continues to extend the new leading (light) strand. Meanwhile a new lagging strand is synthesised in the same way as in chromosomal replication; each Okazaki fragment is primed by DNA primase and extended by the activity of DNA polymerase III. An important feature is that ColE1 replication is **unidirectional** and the replication fork continues around the parental molecule until it returns to the origin. Furthermore, since DNA polymerase I is required to initiate replication, ColE1 is unable to replicate in *polA1* mutants of *E. coli* which lack the DNA polymerase (but not the exonuclease) activity of DNA polymerase I.

There are normally about 15 copies of ColE1 per cell and it appears that this constant copy number is maintained by controlling the rate of plasmid replication in relation to the rate of cell division. In ColE1 another RNA, known as **RNA I**, is transcribed from the light strand commencing from a promoter 445 nucleotides upstream from *oriV* (figure 1.6(1)); RNA I is 108 nucleotides long and has the complementary nucleotide sequence to the 5′ end of RNA II. Thus RNA I can bind to RNA II and alter its secondary structure in such a way that RNase H can no longer cut RNA II at the origin; as a result there is no 3′-OH primer terminus and replication cannot be initiated. The binding of RNA I to RNA II is promoted by a small polypeptide encoded by the plasmid *rop* gene (*repressor of primer*) so that this protein, in conjunction with RNA I, can control the rate of plasmid replication. Whether or not replication is inhibited will depend upon the concentrations of RNA I and the *rop* protein and this, in turn, will depend on the number of copies of ColE1 in the cell—the more copies there are the higher will be the concentrations of RNA I and the *rop* protein and the greater will be the degree of inhibition; as the cell enlarges and divides, these concentrations will drop and the plasmid will replicate until equilibrium is once again reached.

Replication of the large conjugal plasmids is also dependent on host-encoded proteins, but, unlike ColE1, at least one plasmid-encoded protein is essential. Some conjugal plasmids, such as R1 and R100 (section 3.3) replicate unidirectionally while the F plasmid normally replicates bidirectionally. These plasmids, unlike ColE1, use DNA primase to prime replication and DNA polymerase III as the replicase.

The closely related plasmids R1 and R100 both require the protein encoded by the plasmid *repA* gene in order to initiate replication at *oriV* so that the copy number (either one or two) can be controlled by regulating the level of expression of *repA*; this, in turn, is dependent on the *copA* and *copB* genes (*cop*, copy number) and mutations in either of these genes can result in a ten-fold increase in copy number.

The most direct control, also dependent upon an interaction between complementary strands of RNA, operates through the RNA transcribed from *copA*. This RNA (RNA I) is transcribed from the opposite DNA strand to the *repA* messenger (RNA II) and overlaps the 5′ end of the *repA* messenger (figure 1.7); thus RNA I can bind to the base complementary sequence on the *repA* messenger, blocking the ribosome binding sequences and preventing translation of the *repA* protein. As in the ColE1 system of control, the concentration of

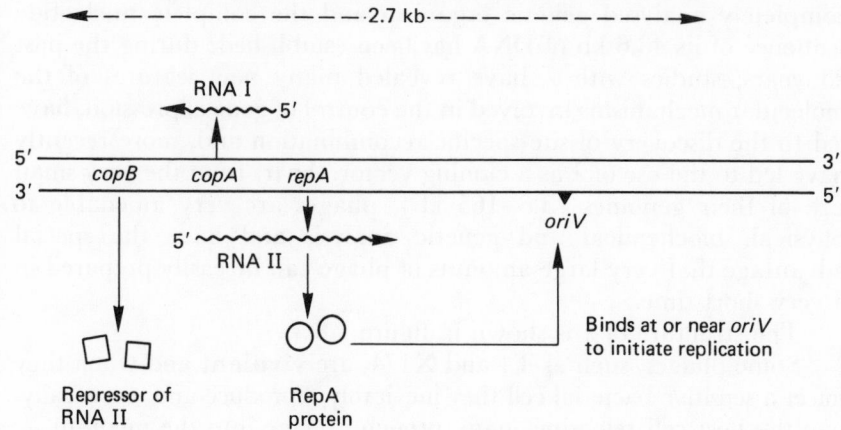

Figure 1.7
The control of replication of the R1 plasmid

Initiation requires the *repA* protein to bind at or near *oriV* and it is regulated by controlling the production of this essential protein. RNA II (the *repA* messenger) and RNA I (the *copA* transcript) are transcribed from opposite strands of DNA and RNA I (80–90 nucleotides long) overlaps the 5′ end of RNA II. It is believed that RNA I interacts with RNA II by complementary base pairing and that this blocks the sequences to which the ribosomes must bind in order to translate *repA*.

The synthesis of the *repA* protein is additionally controlled by the *copA* protein which represses the transcription of RNA II.

The 2.7 kb fragment shown includes all the genetic information required for replication.

RNA I will depend on the number of plasmid copies present. A further control is effected through *copB*; this encodes a repressor protein which can prevent transcription of the *repA* message.

Although R100 is nearly 90 kb long, all but 2 kb in the *oriV* region can be removed without affecting its ability to replicate.

By way of contrast, the control of copy number of the F plasmid does not seem to depend upon an RNA–RNA interaction and at least one F-encoded protein appears to be involved. This protein, known as protein E, is encoded by a gene within the incompatibility (*inc*) region of F (see figure 3.1) and is absolutely essential for the initiation of replication. It is not known how the E protein acts but the current idea is that it has two functional domains which enable it to bind to alternative sequences on F. Firstly, it can bind to an operator sequence at the start of the E gene so that it can regulate its own synthesis. When the E protein is bound to this operator the *E* gene cannot be transcribed but when the concentration of the E protein falls below a certain level the operator is freed from representation and the *E* gene is once again transcribed. Secondly, the E protein can bind to two further sequences in the *inc* region, one at each side of *E*, and this binding positively stimulates the replication of F.

1.5   Bacteriophage λ

Bacteriophages, or phages, are viruses that grow on bacteria and by comparison with bacteria they have a very simple basic structure, consisting of a molecule of DNA (or sometimes RNA) packaged (or encapsidated) into a protein coat, the capsid. The molecule of DNA, the phage chromosome, contains from several to many genes but since these are inactive when within the capsid phages can only reproduce inside a suitable bacterial host cell.

Phages that infect *E. coli* (coliphages) have played a very important role in the development of molecular biology and many of the basic principles of molecular genetics were first established using phage systems. Phage λ is the outstanding example as it is probably the most completely analysed genetic organism and the complete nucleotide sequence of its 48.6 kb of DNA has been established; during the past 25 years, studies with λ have revealed many new features of the molecular mechanisms involved in the control of gene expression, have led to the discovery of site-specific recombination and, more recently have led to the use of λ as a cloning vector. Apart from the very small size of their genomes (3.6–165 kb), phages are very amenable to physical, biochemical and genetic analysis and have the special advantage that very large amounts of phage can be easily prepared in a very short time.

The structure of λ is shown in figure 1.8.

Some phages, such as T4 and X174, are **virulent** and when they infect a sensitive bacterial cell they inevitably reproduce and, eventually, lyse the host cell releasing many progeny phage into the medium. λ, however, is a **temperate** phage and has a more complicated life cycle (figure 1.9). The infective process commences with the tail fibre of a λ particle, recognising and binding to a special receptor site on the bacterial cell surface; this **adsorption** requires the gene product of the *E. coli lamB*$^+$ gene. The tail now penetrates the cell and the phage DNA is injected into the host cell, the empty capsid remaining outside the cell. At this stage several λ genes, known as the **early** genes, are expressed. It is now that λ must decide which of two alternative pathways to follow and if the decision is made to follow the pathway leading to the **lytic** or **vegetative** response (box 1.3), then the **delayed early** genes are expressed and the λ DNA repeatedly replicates (section 1.5.1). After the completion of DNA replication the λ **late** genes are switched on; these encode the 15 or so proteins that are the structural components of the head, tail and tail fibres, and others that are involved in the assembly of these components and the packaging of the DNA into the phage heads. Eventually, the products of two further genes (*S* and *R*) cause the host cell to lyse, releasing about 100 phage progeny into the medium. At 37°C the lytic cycle is completed in 40–50 minutes. The most important lambda genes and genetic control sites are listed in table 1.3

When λ phages infect a population of *E. coli* cells, some of the phages follow the lytic pathway while others follow the **lysogenic response** and **lysogenise** the host cells. In a lysogenic cell, or **lysogen**, a dormant phage genome is retained and replicates synchronously with the bacterial chromosome so that every daughter cell also receives a copy of the phage genome, now referred to as **prophage**. In λ lysogens the phage DNA is normally inserted into the continuity of the bacterial chromosome and is maintained as if it were a group of bacterial genes. The frequency of lysogenisation can vary from zero to nearly unity and which pathway λ follows depends on a complex interaction between genetic and environmental factors. In general, the lytic pathway is favoured when the host cells are growing in a rich environment, while lysogenisation predominates in poorer conditions thus permitting the prophage to survive until environmental conditions once again become favourable for lytic growth.

Icosahedral head

60 nm

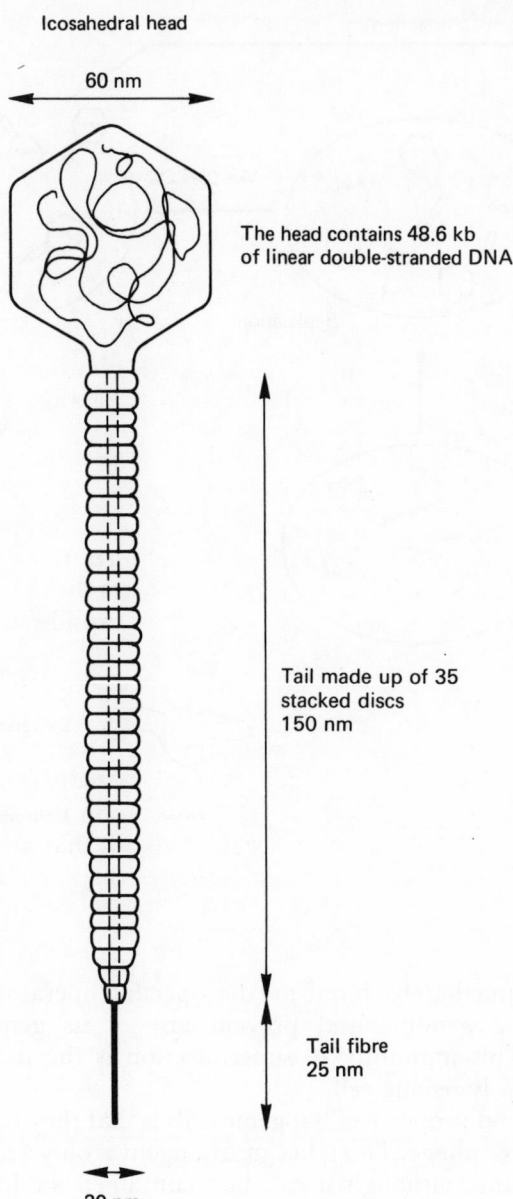

The head contains 48.6 kb
of linear double-stranded DNA

Tail made up of 35
stacked discs
150 nm

Tail fibre
25 nm

20 nm

Figure 1.8
The structure of
bacteriophage λ

The lysogenic state is maintained because one particular λ gene, λcI, remains active during lysogeny. This gene produces a specific repressor which binds to two operator sites on the prophage and prevents any other λ genes from being expressed and so prevents replication. λcI mutants do not produce this repressor and are unable to establish themselves as prophage as they will enter the lytic pathway as soon as they infect a sensitive cell. Lysogenic cells are also **immune** to superinfection; a λ lysogen cannot be lysed or lysogenised by another λ particle because the *cI* repressor present in the cytoplasm of the host

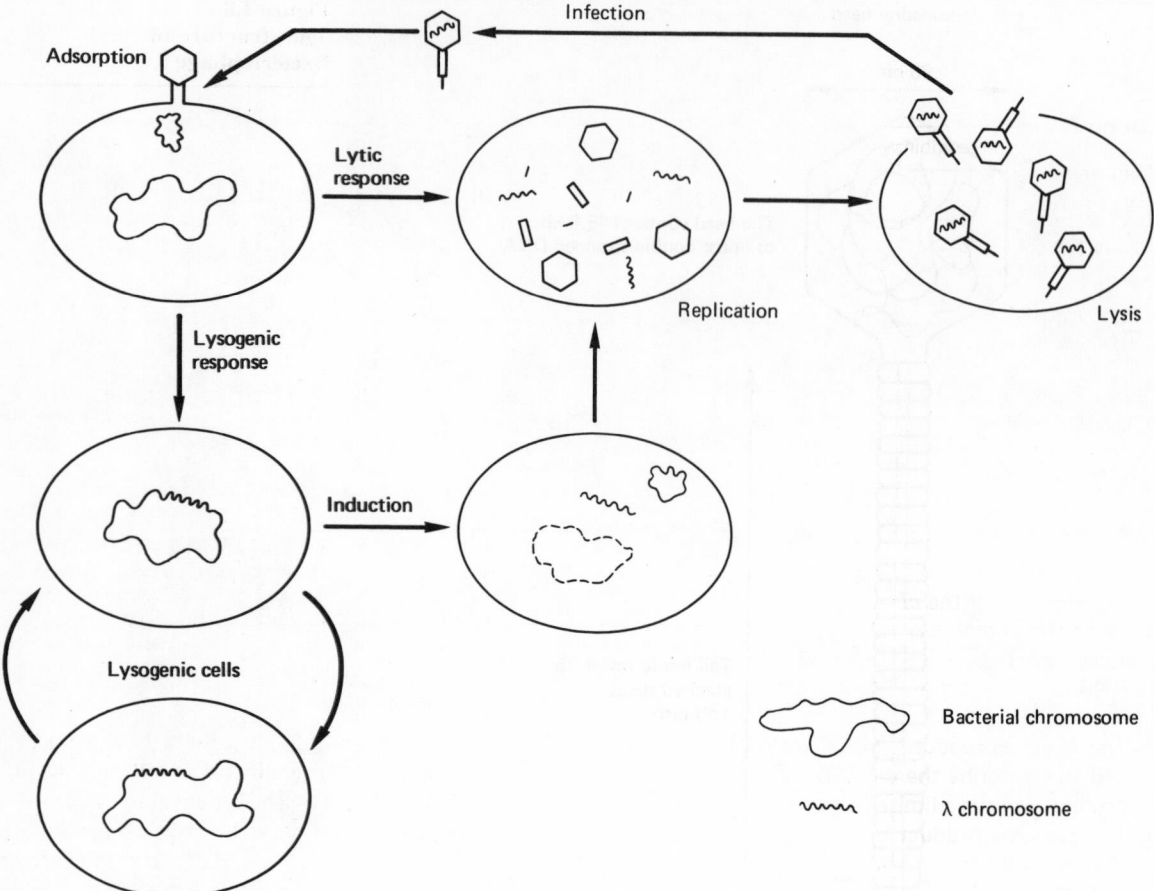

Figure 1.9
The life cycle of phage λ

Upon infection the λ chromosome can either replicate, eventually producing about 100 mature λ particles (the lytic response) or it can integrate into the bacterial chromosome (the lysogenic response) and be inherited in the same way as a group of bacterial genes. This association is not permanent and the prophage can be induced, when it leaves the bacterial chromosome and enters the lytic cycle.

cell will immediately bind to the specific operator sites on the superinfecting genome and prevent any of its genes from being expressed. This immunity to superinfection is the first fundamental property of a lysogenic cell.

The second property of lysogenic cells is that they retain the ability to produce free phage. This is because lysogeny is only a semi-permanent association and, although it can be maintained for hundreds of cell generations, prophage is occasionally released from the bacterial chromosome and enters the lytic cycle. This occurs spontaneously at a low frequency (about 10^{-5} per cell per generation) but can be **induced** in up to 90 per cent of the lysogenic cells by exposing the cells to a low dose of ultraviolet light. Induction occurs because ultraviolet irradiation indirectly destroys the *cI* repressor (see section 6.7.1) and releases the phage genes from repression.

1.5.1 The λ chromosome and its replication

The chromosome of λ is a single molecule of double-stranded DNA 48.6 kb long and it is unusual, as it can exist in two alternative forms.

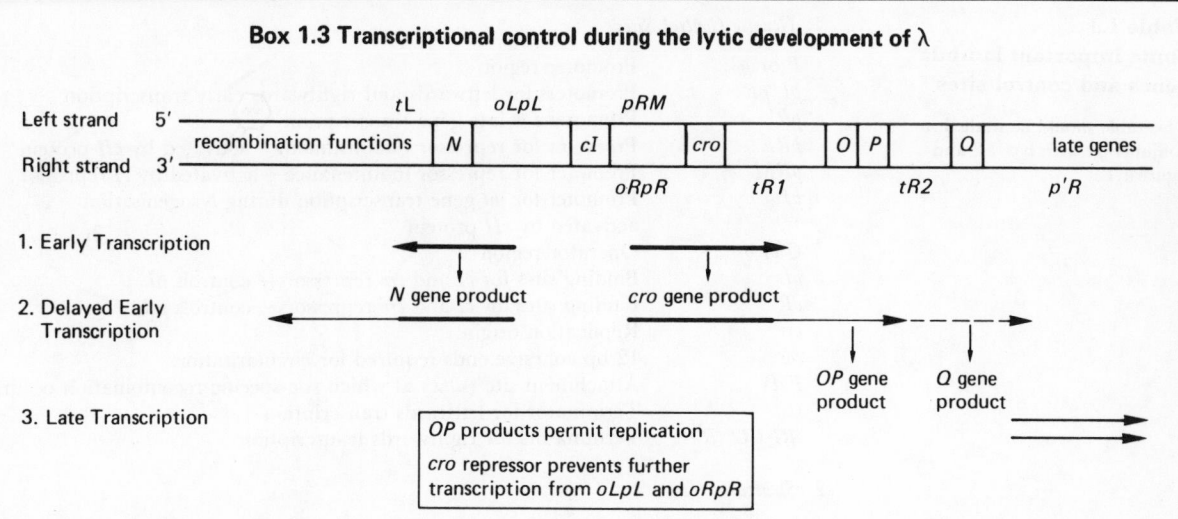

Box 1.3 Transcriptional control during the lytic development of λ

In a lysogenic cell the *cI* gene is transcribed from *pRM* (*p*romoter for *r*epressor *m*aintenance); the *cI* repressor binds to the complex operator–promoter sites *oLpL* and *oRpR* and prevents any other transcription. Induction destroys this repressor and allows the lytic cycle to commence.

1. RNA polymerase binds to *oLpL* and *oRpR* and transcribes the **early** genes as far as *tL* (to the left) and *tR1* (to the right). The *N* and *cro* gene products are made.
2. The *N* gene product is a positive regulatory protein and it allows transcription to read through *tL* and *tR1* and to transcribe the **delayed early** genes. Rightwards transcription normally terminates at *tR2* but the *N* product permits a limited amount of read through to beyond *Q*. The *O, P* and *Q* gene products are made.
3. The *cro* gene product is a further repressor which now binds to *oLpL* and *oRpR* and prevents any further delayed early transcription. The *O* and *P* gene products allow replication to commence and produce many copies of λ DNA.

The *Q* gene product is another positive regulatory protein and it promotes transcription of the **late** genes from *p'R* along each copy of λ DNA.

Note Genetic control sites are shown above and below the lines representing the left and right strands of λ DNA. These strands are designated according to the direction in which they are transcribed.

In mature phage there is a unique molecule of linear DNA and it is in this form that the DNA is injected into the host cell; however, once inside the newly infected cell the DNA circularises. This is possible because the linear molecule has **cohesive** or 'sticky' ends. These 5' single-stranded ends are 12 nucleotides long and they have complementary base sequences; thus the DNA molecule can circularise by pairing between the complementary bases in the cohesive ends (figure 1.10) and DNA ligase can then seal the nicks to form an intact or covalently bonded circular molecule. The ability to circularise is essential for the survival of λ as linear molecules of λ DNA are not only unable to replicate but they cannot be integrated into the host chromosome to establish lysogeny.

The replication of λ is unusual as the normal replicative process

Table 1.3
Some important lambda genes and control sites

This table should be studied in conjunction with box 1.2 and figure 1.12.

1. *Genetic Control Sites*

P or p	Promoter region
pL pR	Promoters for leftwards and rightwards early transcription
pR'	Promoter for late gene transcription
pRE	Promoter for repressor establishment—activated by *cII* protein
pRM	Promoter for repressor maintenance—activated by *cI* repressor
pI	Promoter for *int* gene transcription during lysogenisation—activated by *cII* protein
O or o	Operator region
oL	Binding sites for *cI* and *cro* repressors—controls pL
oR	Binding sites for *cI* and *cro* repressors—controls pR
ori	Replication origin
cos	12 bp cohesive ends required for circularisation
PoP'	Attachment site (*att*λ) at which site-specific recombination occurs
tL	Terminator for leftwards transcription
tR1 tR2 tR'	Terminators for rightwards transcription

2. *Structural Genes*

cI	The λ repressor protein—essential for lysogenisation
cro	*cro* repressor, required for lytic development—prevents *cI* repressor synthesis and, subsequently, turns off all early gene transcription
cII	Positive regulatory protein required for turning on transcription of *cI* (from *pRE*) and *int* (from *pI*) during lysogenisation
cIII	Stabilises the *cII* protein
N	Positive regulatory protein required for all essential processes—permits transcriptional readthrough past *tL*, *tR1* and *tR2*
Q	Positive regulatory protein required for late gene transcription—permits transcriptional readthrough past *tR'*
int xis	Site-specific recombination enzymes
exo	Exonuclease—involved in general recombination
gam	Promotes rolling-circle replication by inactivating the host cell encoded exonuclease V
red	Recombinational protein
O P	Required for theta-type replication
R S	Required for lysis of the cell wall (*R*) and membrane (*S*)
nul A W B C nu3 D E F1 F2	All required for normal heads. *E* encodes the major head protein and *A* the terminase which cuts concatamers at the cohesive ends
Z U V G T H M L K I J	All required for normal tails. *V* encodes the major tail protein and *J* the tail fibres

involves two quite different modes of replication. After the infecting phage genome has circularised it undergoes one or two cycles of θ-mode replication (in exactly the same way as the *E. coli* chromosome replicates (figure 1.3)), but about 10 minutes after infection θ-mode replication ceases and the circular daughter molecules commence to replicate according to the **rolling-circle** mechanism (figure 1.11). It is believed that endonuclease activity makes a single-strand nick in one strand (the (+) strand) of the covalently bonded circular molecule and that the original (+) strand is peeled off at the 5' end as a single-stranded tail. As the (+) strand is peeled off at the 5' end, so DNA polymerase

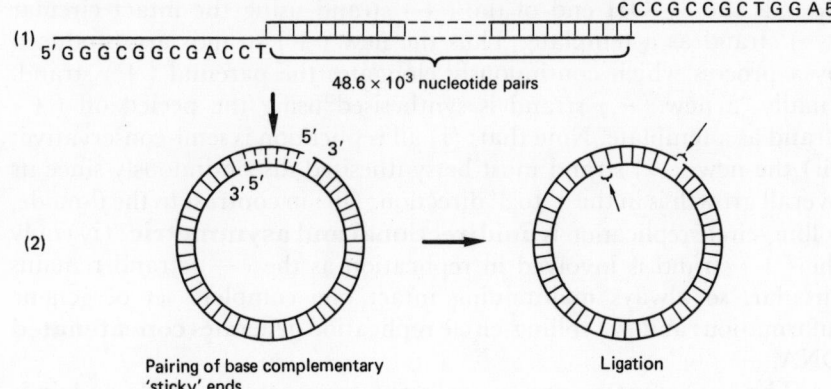

Figure 1.10
The cohesive ends of the λ chromosome

The single-stranded base complementary ends of the λ chromosome (1) can anneal by the formation of H-bonds between the complementary bases to form a circular duplex of DNA (2), and DNA ligase can seal the gaps to produce a covalently closed circular molecule. The actual sequence of the cohesive ends is shown in (1).

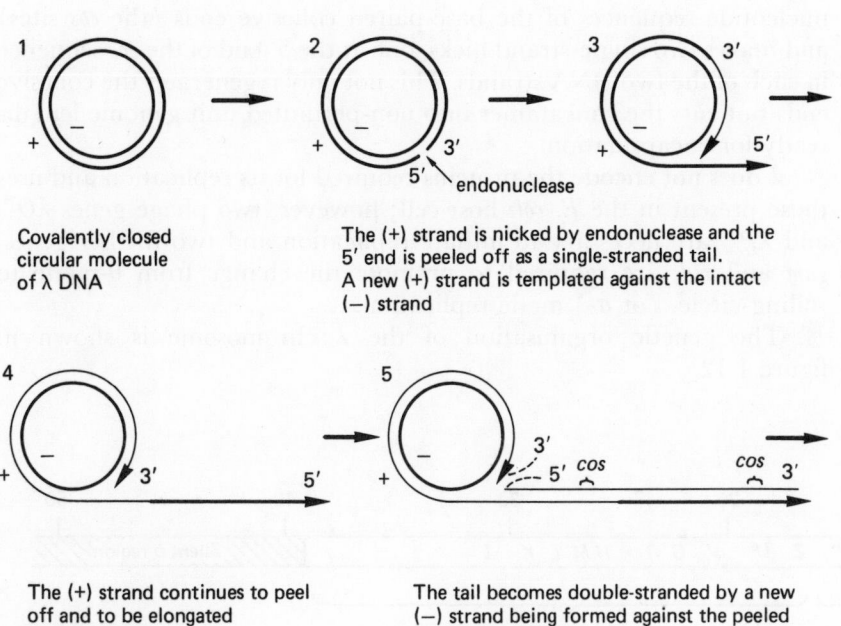

1 Covalently closed circular molecule of λ DNA

2 The (+) strand is nicked by endonuclease and the 5' end is peeled off as a single-stranded tail. A new (+) strand is templated against the intact (−) strand

3

4 The (+) strand continues to peel off and to be elongated continuously

5 The tail becomes double-stranded by a new (−) strand being formed against the peeled off (+) strand

6 The Ter enzyme recognises the *cos* sequences along the concatamer and makes staggered nicks. This cuts the concatamer into unit genome lengths and regenerates the cohesive ends

Figure 1.11
Rolling-circle replication of the λ chromosome

When λ enters the lytic cycle the chromosome undergoes one or two cycles of θ-mode replication. Each of these progeny molecules now becomes a template for rolling-circle replication.

extends the 3′–OH end of the (+) strand using the intact circular (−) strand as a template. Thus the new (+) strands are produced by a process which continuously elongates the parental (+) strand. Finally, a new (−) strand is synthesised using the peeled off (+) strand as a template. Note that: (i) all replication is semi-conservative; (ii) the new (−) strand must be synthesised discontinuously since its overall growth is in the 5′ to 3′ direction; (iii) in contrast to the θ-mode, rolling-circle replication is **unidirectional and asymmetric**; (iv) only the (+) strand is involved in replication as the (−) strand remains circular, so always maintaining intact one complete set of genetic information; and (v) rolling-circle replication generates **concatenated** DNA.

The concatenates are up to eight genomes in length and it is important to note that these are a direct product of replication and not, like the concatenates found after T2 or T4 infection, a product of recombination. The concatenated DNA must now be packaged into phage heads. Unlike some phages λ does not package a 'headful' of DNA and, instead, a special endonuclease known as Ter (for *ter*minus generation), the product of the λ *A* gene, recognises the specific nucleotide sequences of the base-paired cohesive ends (the *cos* sites) and makes two single-strand nicks, one at the 5′ end of the *cos* sequence in each of the two DNA strands. This not only regenerates the cohesive ends but cuts the concatamer into non-permuted unit genome lengths ready for encapsidation.

λ does not encode the proteins required for its replication and uses those present in the *E. coli* host cell; however, two phage genes λO⁺ and λP⁺ are necessary to initiate replication and two further genes, *gam* and *red*, are required to promote the change from θ-mode to rolling-circle- (or σ-) mode replication.

The genetic organisation of the λ chromosome is shown in figure 1.12.

Figure 1.12
The genetic map of λ

The map shows the sequences of some of the principal genes along the vegetative chromosome of λ. Note the tendency for genes of related function to be clustered together.

The map distances are expressed as a percentage of the 48.6 kb that make up the phage chromosome. Structural genes are shown between the two strands of DNA and genetic control sites either above the left strand (leftwards transcription) or below the right strand (rightwards transcription).

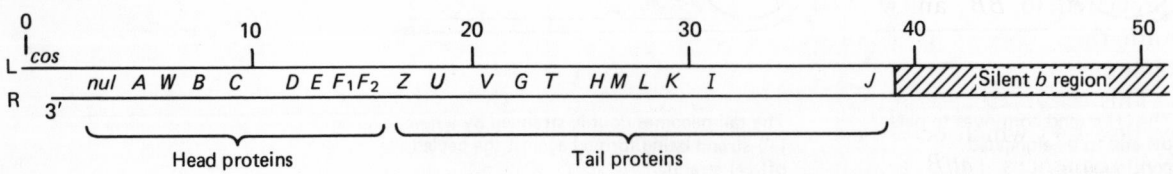

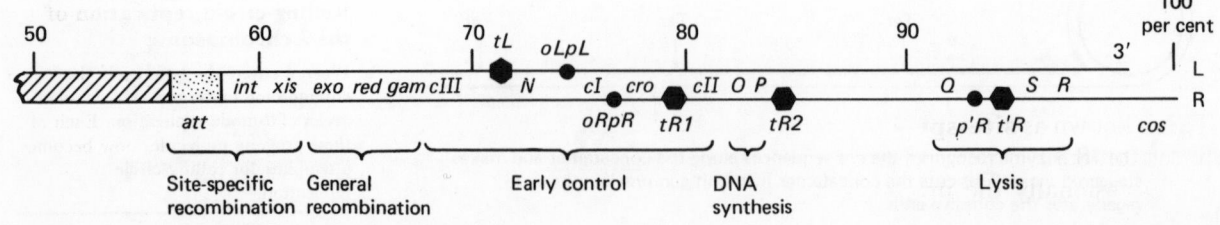

1.5.2 Prophage integration and site-specific recombination

λ was discovered by **Esther Lederberg** in 1951 and during the next decade many mutants were isolated and mapped. At first it was thought that in a lysogen the prophage was inherited as an extra-chromosomal element like a plasmid, but it soon became clear that the prophage was in some way associated with the *E. coli* chromosome. For the next 10 years there was considerable disagreement as to the nature of this association; one school argued that prophage was simply attached to the bacterial chromosome and another considered that prophage was inserted into the linear continuity of the chromosome. It was at the height of this controversy, in 1962, that **Allan Campbell** proposed an elegant model to explain how prophage integration could occur. Although there is now very convincing evidence in support of the model it is interesting to note that Campbell, when he first proposed the model, did not seriously believe it to be correct; he suggested the model more because it made several precise predictions which could be proved or disproved experimentally.

The key to prophage integration is that a reciprocal recombination event occurs between a special **attachment site** on the circularised λ chromosome (*attP*) and a corresponding but not identical site on the circular bacterial chromosome (*attB*); the result is that the phage chromosome is inserted into the continuity of the bacterial chromosome (figure 1.13). A requirement of this model is that the phage chromosome must be circular at the time of integration, but when Campbell proposed the model it was thought that the λ chromosome was always linear; however, the following year the λ chromosome was shown to have cohesive ends which could enable it to circularise.

Each *att* site consists of three components. Firstly, a 15 bp core sequence common to all *att* sites and symbolised (o); the actual recombination event or cross-over occurs within and between these core sequences. Secondly, a pair of dissimilar flanking sequences, B and B' on the bacterial chromosome and P and P' on the λ chromosome. Thus *attB* and *attP* can be represented as BoB' and PoP' (usually abbreviated to BB' and PP'). Because crossing-over always occurs within these core sequences, integration generates two new hybrid attachment sites, known as *attL* (or BoP') and *attR* (or PoB').

This event involves a very special type of genetic recombination (see box 1.4) which occurs between two different but highly specific DNA sequences (*attB* and *attP*). The integration of λ is $recA^+$-independent, and so can occur in $recA^-$ hosts, but it is absolutely dependent upon the λint^+ gene product, the **integrase** protein, and the overall reaction can be written

$$BoB' + PoP' \xrightarrow{\text{integrase}} BoP' + PoB'$$

This is known as **site-specific recombination** and is further discussed in section 8.3.

Occasionally the prophage can leave the bacterial chromosome,

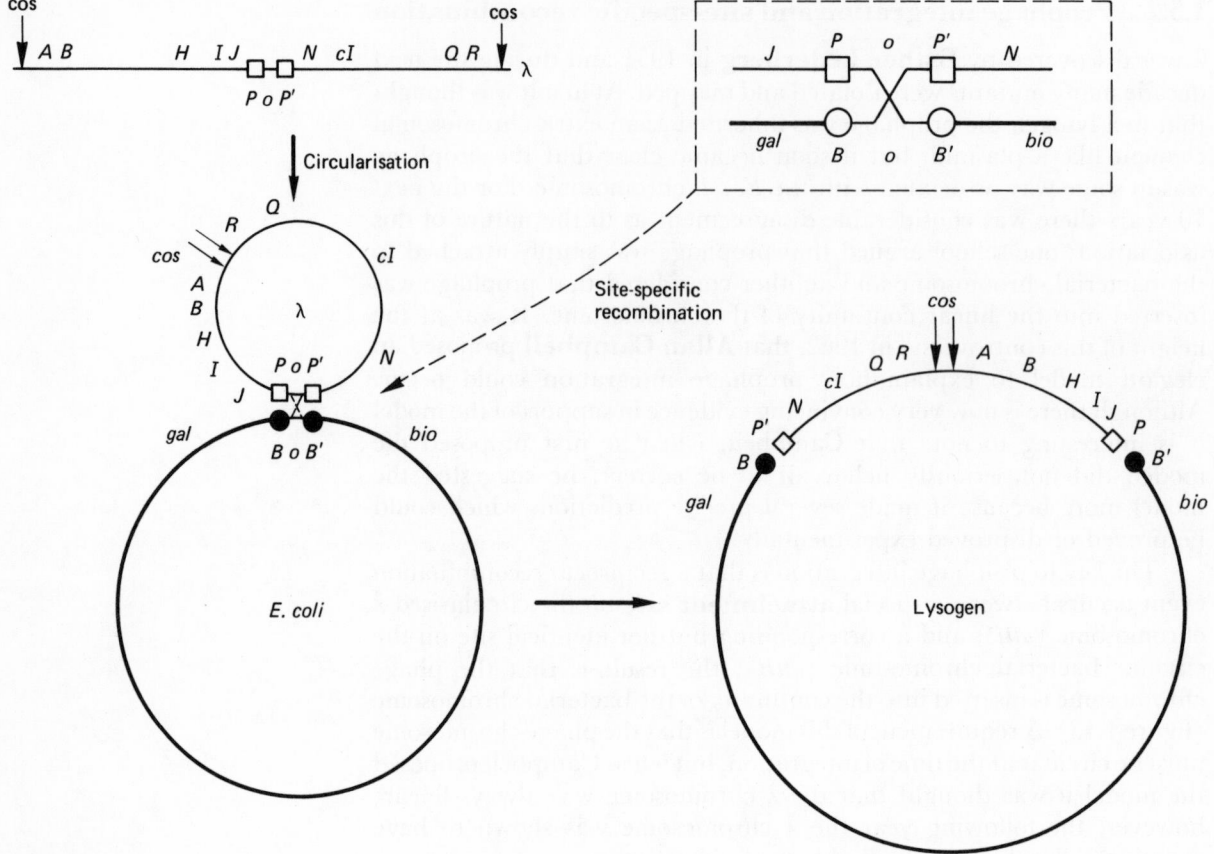

Figure 1.13
Lysogeny

A single reciprocal cross-over between
PoP' (*attP*) on the λ chromosome and
BoB' (*attB*) on the bacterial
chromosome integrates λ into the
bacterial chromosome.

either spontaneously or after induction, and this **prophage excision**
occurs by the reverse process (figure 1.14). This involves a site-specific
recombination event between the hybrid attachment sites *BoP'* and
PoB'; this requires not only the integrase protein but also the product
of the λ*xis*[+] gene, sometimes referred to as excisionase, and the overall
reaction can be written

$$BoP' + PoB' \xrightarrow{\text{integrase \& excisionase}} BoB' + PoP'$$

Under normal conditions λ only integrates at *attB*, but if no functional
attB site is present then λ can be forced to integrate at other sites around
the *E. coli* chromosome; these secondary attachment sites are known
to have nucleotide sequences resembling the sequence of *attB*.

A further important point is that the vegetative chromosome of λ
and the prophage have different gene orders—they are cyclic
permutations of each other (see figure 1.13). This is because *attP* is not
located within the cohesive end sequences of the vegetative chromosome.

Box 1.4 Types of recombination

1. General Recombination

Also known as homologous or $recA^+$-dependent recombination, it is responsible for re-assorting the genetic material, thus creating new linkage relationships between genes.

It occurs between two nucleotide sequences that are wholly or largely identical. These **homologous** sequences are usually two different but equivalent DNA molecules, such as the chromosomes from two different strains of *E. coli*. However, there may be quite long regions of homology on two very different molecules or even within the same molecule; thus both the F plasmid and the *E. coli* chromosome carry copies of the insertion sequence IS2 and these provide homologous sequences between which general recombination can occur. General recombination is totally dependent on the *E. coli* $recA^+$ gene while the major recombinational pathway (chapter 5) is also dependent on the $recB^+$ $recC^+$ and $recD^+$ genes. These genes encode proteins involved in general recombination.

2. Site-specific Recombination

Site-specific recombination is a special type of reciprocal recombination that only occurs between two special recombination sites, one of which must be present on each participating molecule; these pairs of sites, each one of which is a short and very specific nucleotide sequence, are recognised by the special proteins that promote site-specific recombination. Unlike general recombination, site-specific recombination does not require the $recA^+$ gene. There are two types of site-specific recombination:

(i) Conservative. This is non-replicative and is best illustrated by the integration and excision of λ prophage.
(ii) Replicative. This initiates the transposition of certain transposable elements and requires specific transposase proteins (section 7.4.2); the element transposes to a new location without being lost from its original site.

3. Illegitimate Recombination

Also known as illegitimate crossing-over, this requires neither regions of extensive homology nor specific sequences. Almost nothing is known about the process (or processes) but in some instances it is known to be promoted by insertion sequences.

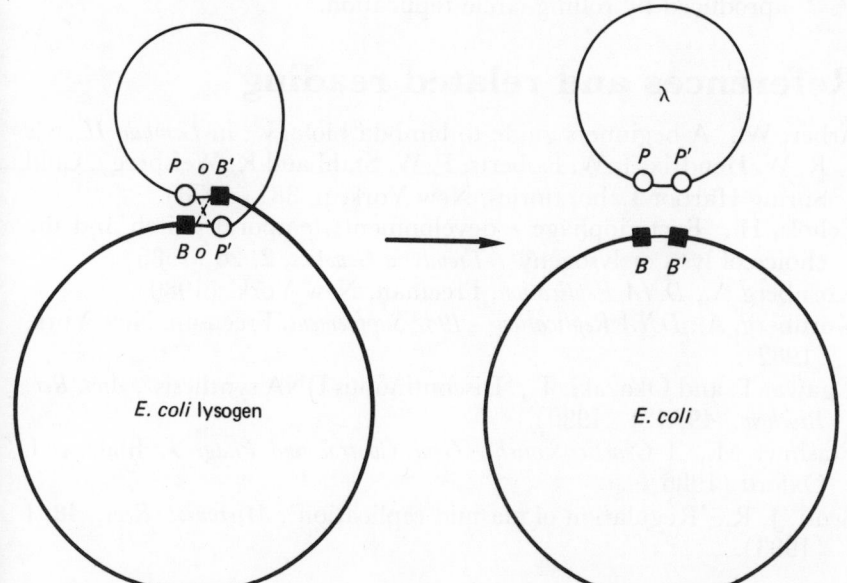

Figure 1.14
Phage excision

A single cross-over between *BoP'* (*attL*) and *PoB'* (*attR*) excises the λ prophage from the chromosome of a lysogen.

Exercises

1.1 Compare and contrast the θ and rolling-circle modes of replication.

1.2 How does a replication fork move unidirectionally along a replicating DNA molecule when all known DNA polymerases can only add nucleotides on to the $3'$ end of a growing polynucleotide chain?

1.3 Explain how replication is initiated, bearing in mind that DNA polymerase can only add nucleotides on to an **existing** polynucleotide strand.

1.4 What are the functions of the different proteins found associated with the replication fork?

1.5 How does *E. coli* cope with the occasional mismatched nucleotides that are introduced during replication?

1.6 The bacterial cell may contain, in addition to its chromosome, other molecules or sequences of DNA. What are these and what roles do they play?

1.7 Distinguish between conjugative and non-conjugative plasmids.

1.8 How may RNA control the number of copies of a particular plasmid present in a cell?

1.9 Compare replication of the F and ColE1 plasmids.

1.10 λ is a temperate phage. Distinguish between the lytic and lysogenic responses.

1.11 What is site-specific recombination and how does it differ from general or homologous recombination? How does it result in the integration of λ prophage?

1.12 Vegetative λ and λ prophage have different but circularly permuted genetic maps. Explain this.

1.13 What is concatenated DNA? How are unit length λ genomes produced by rolling-circle replication.

References and related reading

Arber, W., 'A beginners guide to lambda biology', in *Lambda II* (eds R. W. Hendrix, J. W. Roberts, F. W. Stahl and R. Weisberg), Cold Spring Harbor Laboratories, New York, p. 381 (1983).

Echols, H., 'Bacteriophage λ development; temporal switch and the choice of lysis or lysogeny', *Trends in Genetics*, **2**, 26 (1986).

Kornberg A., *DNA Replication*, Freeman, New York, (1980).

Kornberg, A., *DNA Replication—1982 Supplement*, Freeman, New York, (1982).

Ogawa, T. and Okazaki, T., 'Discontinuous DNA synthesis', *Ann. Rev. Biochem.*, **49**, 421 (1980).

Ptashne, M., *A Genetic Switch—Gene Control and Phage λ*, Blackwell, Oxford (1986).

Scott, J. R., 'Regulation of plasmid replication', *Microbiol. Revs.*, **48**, 1 (1983).

TRANSDUCTION AND 2
TRANSFORMATION

2.1 Introduction

Transduction was discovered by **Joshua Lederberg and Norton Zinder** in 1951, while looking for a system of conjugation and recombination (chapter 4) in *Salmonella typhimurium*. When they grew together two different mutant strains, one of which was lysogenic for phage 22 (P22), they were able to recover rare prototrophs from the mixed culture. They were surprised to find that cell contact was not required since prototrophs were still recovered when the two strains were grown side by side but separated by a fine sintered glass filter through which the bacteria could not pass, and they concluded that this genetic exchange was promoted by a DNase-resistant filterable agent. Very soon afterwards Zinder demonstrated that the filterable agent was the temperate phage P22 and that the phage particles themselves were capable of transferring fragments of genetic material from a donor to a recipient strain, where they could be recombined into the recipient chromosome. They called this process **transduction**, literally 'leading across'.

2.2 Generalised transduction

Transduction, the transfer of a small piece of DNA from one bacterial cell to another by a phage particle, has been described in many different bacteria but it has been most extensively used to analyse the genetic fine structure of the *S. typhimurium*, *E. coli* and *B. subtilis* chromosomes; it is also extensively used as a tool in strain construction. We will primarily consider P1-mediated transduction in *E. coli*, discovered by **E. S. Lennox** in 1955; this, like PSS-mediated transduction in *Salmonella*, is a system of **generalised** transduction, so-called because almost any bacterial gene can be transferred from a donor to a recipient.

P1 has an icosahedral head enclosing a single linear molecule of DNA approximately 100 kb long and a complex tail made up of a rigid core tube, a contractile sheath, base plate and tail fibres. It is a temperate phage and so, when it infects an *E. coli* cell, it can either enter the lytic cycle and lyse the cell or it may lysogenise it. When P1 lysogenises a cell it does not, like many temperate phages (λ, for example) integrate into the chromosome of the host cell but the phage genome is maintained as a circular molecule of autonomously replicating plasmid DNA. However, in transduction experiments it is customary to use the virulent mutant P1$_{vir}$ which does not lysogenise; this has the advantage that the transductants are not lysogenised by the normal phage particles present in the transducing lysate and so remain phage-sensitive and can be used in further transduction experiments.

In the laboratory, transduction is carried out in two stages. First, a cell-free suspension of phage is prepared by growing P1 on a suitable donor strain of *E. coli* and, second, these phages are used to infect a genetically different recipient strain and genetic recombinants (called **transductants**) selected by plating the infected cells on a suitable selective medium.

P1-transducing phage is prepared by the lytic infection of 10^8 donor bacteria with about 10^6 P1 particles; after the phage has replicated and the donor cells have lysed, any bacterial debris is removed by centrifugation and the supernatant is shaken with chloroform to kill any surviving bacteria. The vast majority of the maturing phage will encapsidate a headful (about 100 kb) of P1 DNA and these will become normal P1 particles. However, during the later stages of the infective process a small proportion (3×10^{-3}) of the maturing phages encapsidate a segment of the host chromosome instead of packaging a phage genome; these segments, like the phage chromosome, are about 100 kb long, or about 1/40 of the bacterial chromosome. Although the fragments of *E. coli* DNA packaged by P1 are of constant size, the fragments including any particular donor gene are of variable composition; this is because the headful packaging mechanism does not always commence from the same site on the donor chromosome. This is in contrast to the fragments of *Salmonella* DNA packaged by P22 which are of both constant size and composition (see section 2.3.1); it appears that the P22 packaging mechanism always commences from one or more particular sites and then proceeds by packaging sequential headfuls.

These transducing phage particles contain no phage DNA at all and so can neither lyse nor lysogenise the host cells they infect. A good phage preparation may contain as many as 10^{11} particles per ml but only about 0.3 per cent of these contain bacterial DNA and are transducing particles; thus for any given gene the frequency of transducing particles will be about $1/40 \times 3/1000$ or 7.5×10^{-5} (note, however, that some mutants of P1 produce a much higher frequency of transducing phage). Since over 1100 loci have been identified around the *E. coli* chromosome, the average fragment of transducing DNA contains 25–30 known loci.

The phage preparation is now used to infect a population of recipient cells and the mixture plated on selective medium. With $P1_{vir}$ it is necessary either to use a very low multiplicity of infection (0.01) in order to avoid the excess killing of recipient cells or, alternatively, to add sodium citrate to the transduction mixture—this prevents the re-adsorption of any phage produced by the lysis of infected cells and enables the use of multiplicities of infection of up to 1 (**multiplicity of infection** or MOI is the ratio of phage to bacteria).

Figure 2.1 illustrates transduction when P1 grown on a wild type donor (*leu*$^+$) is used to infect a leucine-requiring recipient (*leu*$^-$) and wild type transductants selected by plating on minimal medium. After 2–3 days, many leucine-independent colonies are found growing on the plates. While most of these colonies are probably transductants some, or all, could be due to back-mutations of the *leu*$^-$ recipients to *leu*$^+$; it is important to estimate how many of these colonies are due to spontaneous mutation by plating the recipient bacteria alone on the selective medium—any colonies in excess of the number found on these control plates are almost certainly due to transduction. The formula for this transduction is written

$$leu^- \quad (\times) \quad leu^+$$

with the genotype of the recipient strain on the left and separated from the genotype of the donor by (×), the symbol for transduction.

General transduction is a rare event and only about one transductant is recovered for every 10^5 phage particles used. Thus the simultaneous

Figure 2.1
P1-mediated transduction in *E. coli*

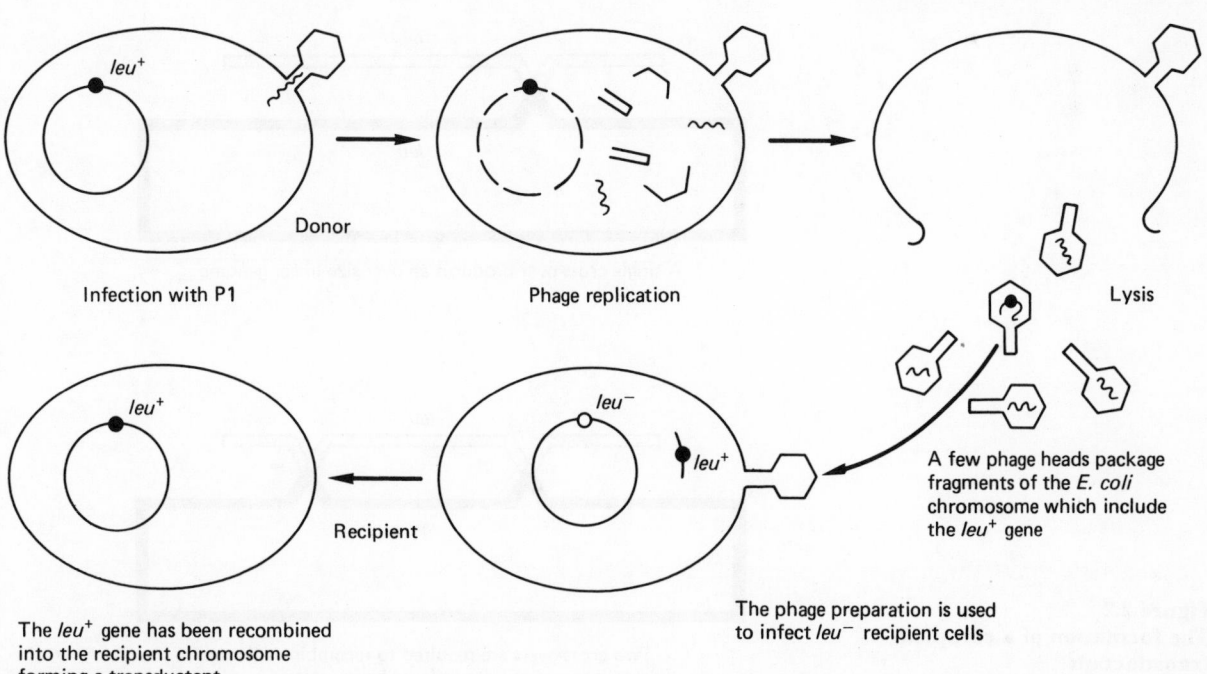

Infection with P1

Phage replication

Donor

Lysis

A few phage heads package fragments of the *E. coli* chromosome which include the *leu*$^+$ gene

The phage preparation is used to infect *leu*$^-$ recipient cells

Recipient

The *leu*$^+$ gene has been recombined into the recipient chromosome forming a transductant transduction of 2 genes on **different** fragments of transducing DNA is very unlikely, having a probability of $10^{-5} \times 10^{-5}$ or 10^{-10}; if two genes are consistently transduced together they must be very closely linked on the bacterial chromosome and so capable of being carried on the same chromosome fragment.

Once the fragment of transducing DNA has been introduced into a recipient cell one of three events must occur:

(i)   The fragment is completely degraded by nuclease activity; only a small proportion of the fragments is thought to be degraded in this way.

(ii)   About 1 in 10 of the fragments undergoes general recombination with the homologous region of the recipient chromosome to produce a stable **complete** transductant with an intact circular chromosome. Since the fragment is linear and the recipient chromosome is circular there must be an even number of cross-overs as otherwise recombination would produce an over-size linear chromosome (figure 2.2); such a structure would be unable to replicate and would rapidly be degraded by exonuclease activity. It is likely that the remaining parts of the fragment, together with the replaced segments of host DNA, are degraded.

(iii) About 90 per cent of the fragments persist in the cytoplasm of the recipient cell without either recombining with the host genome or being degraded, forming what are known as abortively transduced cells. These fragments of DNA, like all other molecules of transducing DNA, are exonuclease resistant

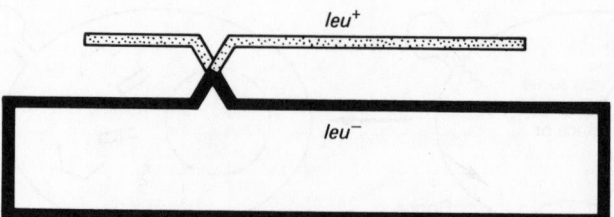

A single cross-over produces an over-size linear genome

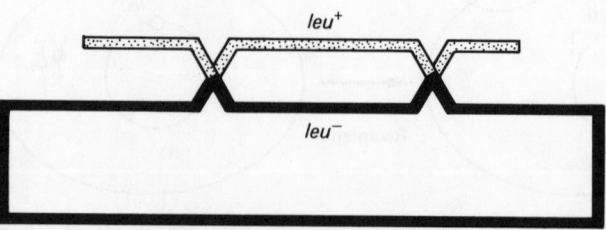

**Figure 2.2
The formation of a complete
transductant**

Two cross-overs are required to recombine a donor
gene into a circular recipient chromosome

apparently because they adopt a circular configuration with the two ends of the DNA duplex being held together by a DNA-associated protein; note that these molecules are **not** covalently closed circular molecules and in other respects will behave as linear molecules.

Since these fragments are unable to replicate, each time the cell divides the donor fragment is transmitted to only one of the two daughter cells; this is known as **unilinear** inheritance and, in effect, the fragment is passed along a single cell line. However, the genes on the fragment may retain the ability to be transcribed and translated so that any wild type alleles may continue to produce functional gene products; in this way a defect in a chromosomal gene can be **complemented** by the corresponding wild type gene on the fragment thus enabling the abortively transduced cell to divide. Each of the daughter cells that does not inherit a fragment may contain a residual amount of the missing gene product, enough to enable one or two further cell divisions to occur, but it will be unable to make more of this gene product. In certain instances the products of these divisions, together with the single cell containing the fragment, may form a minute colony known as an **abortive transductant** (figure 2.3).

In the P1-*E. coli* system, abortive transductants are not normally observed and it would seem that there are very specific requirements for their detection; for example, abortive transductants have been observed in *araC⁻* (×) *araC⁺* transductions but not in the corresponding transductions using *araB⁻* or *araA⁻* recipients. However, in the P22-*Salmonella* system they are frequently observed, particularly if the

Figure 2.3
Abortive transduction

A microcolony, known as an abortive transductant, may result when, for example, an *araC⁻* recipient receives a fragment of donor chromosome carrying an *araC⁺* gene which then fails to be recombined into the recipient chromosome.

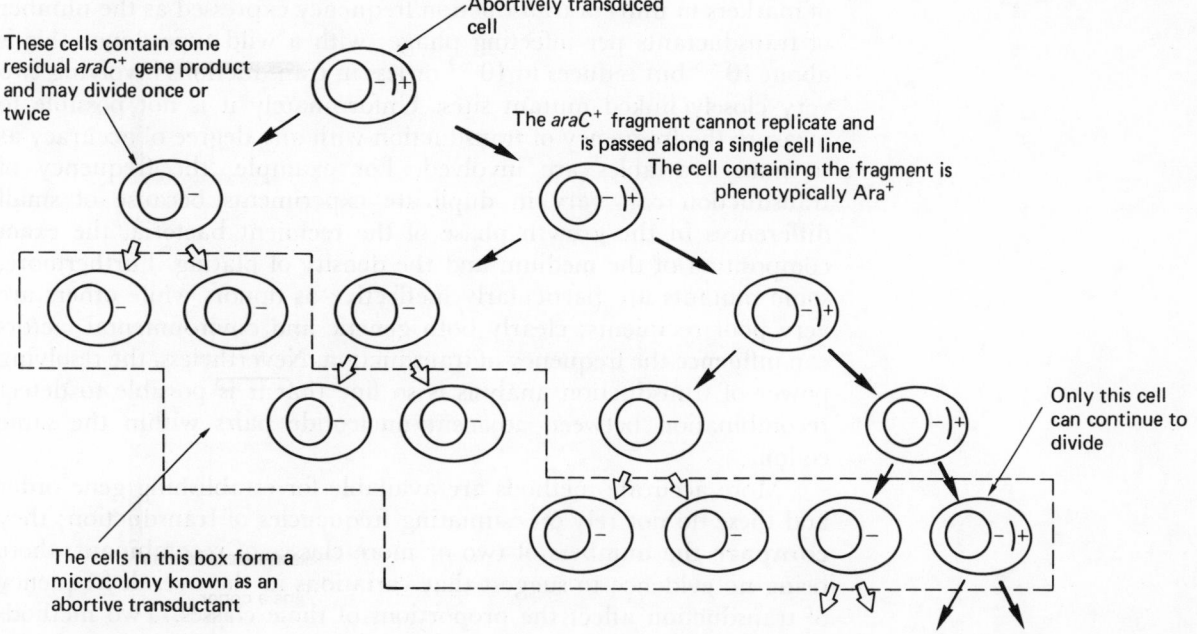

Abortively transduced cell

These cells contain some residual *araC⁺* gene product and may divide once or twice

The *araC⁺* fragment cannot replicate and is passed along a single cell line. The cell containing the fragment is phenotypically Ara⁺

Only this cell can continue to divide

The cells in this box form a microcolony known as an abortive transductant transduction mixture is plated on a medium which restricts the residual growth of the recipient bacteria. In this system, abortive transduction is frequently used as a **complementation test** (section 4.6); if, for example, abortive transductants are observed in a transduction between two different leucine mutants, it is most likely that the two mutations involved are in **different** genes and that the transduction is of the type *leuA⁻ leuB⁺* (×) *leuA⁺ leuB⁻*.

2.3   Mapping by transduction

In higher organisms the genetic distance between two markers is measured in units of percentage recombination, calculated by expressing the total number of recombinants as a percentage of the total number of zygotes. However, in bacterial systems recombination is usually a rare event and it is necessary to use a selective technique to identify just **one** particular class of recombinant, all other classes being excluded from the analysis. For example, in the transduction *leuA⁺ leuB⁻* (×) *leuA⁻ leuB⁺* selection must be made for leucine-independent recombinants and these will all be *leuA⁺ leuB⁺*; the *leuA⁻ leuB⁻* reciprocal recombinants are not recovered as they are indistinguishable from the non-transduced recipient cells. Furthermore, it is not possible to estimate the number of non-recombinant zygotes since these also are indistinguishable from the recipient cells. For these reasons, it is not usually possible to express genetic distances in terms of percentage recombination.

Since the frequency of crossing-over is proportional to genetic distance (an assumption that is made in all mapping experiments), it is theoretically possible to measure the **relative** distances between pairs of markers in units of transduction frequency expressed as the number of transductants per infecting phage; with a wild type donor this is about 10^{-5} but reduces to 10^{-7} or less in transductions involving two very closely linked mutant sites. Unfortunately it is not possible to measure the frequency of transduction with any degree of accuracy as so many variables are involved. For example, the frequency of transduction can vary in duplicate experiments because of small differences in the growth phase of the recipient bacteria, the exact composition of the medium and the density of plating. Furthermore, some mutants are particularly ineffective as donors while others are very poor recipients; clearly both genetic and environmental factors can influence the frequency of transduction. Nevertheless, the resolving power of transduction analysis is so fine that it is possible to detect recombination between adjacent nucleotide pairs within the same codon.

More accurate methods are available for establishing gene order and these do not rely on estimating frequencies of transduction; they **compare** the numbers of two or more classes of recombinant, there being no evidence to suggest that variations in the overall frequency of transduction affect the proportions of these classes. Two methods are commonly used, the **co-transduction method** and the **three-point test cross**.

2.3.1 The co-transduction method

In P1-mediated transduction each transduced fragment is only about 2.5 per cent of the bacterial chromosome and if two markers are consistently co-transduced they must be located within the same fragment of donor DNA and so be very closely linked. Conversely, if they cannot be co-transduced they are probably at least 100 kb apart (the length of a fragment). In the P1 system these fragments are of constant size but, since the headful packaging mechanism does not always commence from the same site on the donor DNA, they are of variable composition. Thus if A^+–B^+ and B^+–C^+ are each separated by 60 kb of DNA, it is possible to co-transduce A^+ with B^+ and B^+ with C^+ but **not** A^+ with C^+:

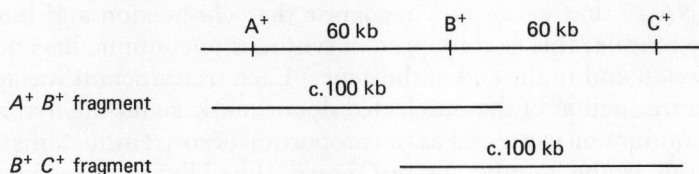

This is in contrast to the P22–*S. typhimurium* system where the packaging mechanism appears to commence from one or more fixed sites so that the transducing fragments do not overlap and have a constant composition; a further difference is that P22 is smaller than P1 and can only package about 42 kb of DNA.

With this method the transductions are of the type A^-B^- ($\times$) A^+B^+ (or A^-B^+ ($\times$) A^+B^-) and A^+ transductants are selected by plating on a medium which permits the unrestricted growth of both A^+B^+ and A^+B^- transductants (figure 2.4). Some of these A^+ transductants are also B^+ (the co-transductants) and their proportion is estimated by testing each A^+ transductant, either by individually streaking or by replica plating, on a medium which distinguishes the B^+ and B^- phenotypes. In order to generate an A^+ transductant there must be an **obligatory** cross-over in interval I, but the second cross-over can be in either interval II or interval III, producing A^+B^- or A^+B^+ transductants respectively. Since the frequency of crossing-over is proportional to genetic distance, it follows that the closer A is to B, the smaller will be the length of interval II, and the fewer the number of cross-overs that will occur within it; hence the proportion of A^+B^+ co-transductants due to crossing-over in intervals I and III

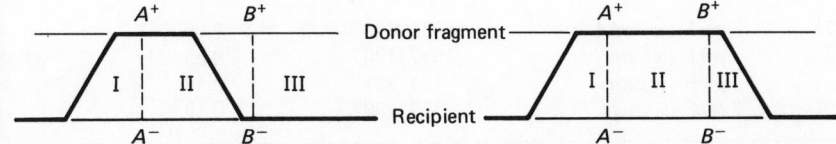

Figure 2.4
Co-transduction

In the transduction A^-B^- ($\times$) A^+B^+, A^+ transductants are selected by plating on minimal medium containing any supplements required for the growth of the A^+B^- recombinants. There is an **obligatory** cross-over in interval I but the second cross-over can occur in either interval II, producing an A^+B^- transductant, or interval III, producing an A^+B^+ co-transductant. A^+ and B^+ are the selected and unselected markers respectively.

Note that the transduced fragment is always shown above and shorter than the recipient chromosome.

will be greater, and the nearer A is to B then the higher will be the proportion of co-transductants. Note that when two markers are co-transducible the proportion of co-transductants depends upon the recombination process and not on the frequency of joint transfer.

As an example of co-transduction analysis, consider the results of the eight transductions set out in table 2.1; these are part of a much larger body of data collected in 1965 by **Ethan Signer, Jonathan Beckwith and Sydney Brenner** while mapping mutations that suppress amber and ochre mutations. In the first set of experiments, selection was made for the transduction of $supC^+$ by using $lacZ2$ as recipient; $lacZ2$ is an ochre mutation suppressible by the dominant $supC^+$ ochre suppressor, so that $lacZ2\ supC^+$ transductants have a Lac$^+$ phenotype. (Note that ochre mutations are chain termination triplets produced by mutation within a gene and they cause the premature termination of protein synthesis. The wild type $supC^-$ allele encodes a minor species of tyrosine tRNA and the $supC^+$ mutation changes the anticodon of this tRNA so that it can now recognise the ochre codon and insert a tyrosine residue; this enables protein synthesis to continue beyond the ochre codon and to the end of the gene.) Each transductant was tested for co-transduction of the unselected donor marker and the frequency of co-transduction expressed as the proportion of co-transductants; thus in (i) this is the number of $supC^+pyrF$ (Lac$^+$Pyr$^-$) transductants expressed as a proportion of the total number of $supC^+$ (Lac$^+$Pyr$^-$ + Lac$^+$Pyr$^+$) transductants. The second set of experiments was similar except that $pyrF^+$ was the selected marker.

From (i) and (iii) it is clear that $trpC$ (0.48) is closer to $supC$ than is $pyrF$ (0.13), but this does not distinguish the alternative gene orders

$$pyrF \ldots supC \ldots trpC \quad \text{and} \quad supC \ldots trpC \ldots pyrF$$

But, from (vi) and (viii) it is clear that $trpC$ (0.55) is closer to $pyrF$ than is $supC$ (0.13); these results are only consistent with the order

Table 2.1
Co-transduction

$supC^+$ is an ochre suppressor and its transduction was selected by using a suppressible $lacZ$ ochre mutant as a recipient. $supC^+$ is a mutation in the gene for tyrosine tRNA$_1$ and is alternatively designated $tyrT$.

The ton^r mutation confers resistance to phage T1. $trpA$ and $trpC$ are both tryptophan-requiring while $pyrF$ requires either thymine or cytosine.

The data are adapted from Signer, E. R., Beckwith, J. R. and Brenner, S., *J. Molec. Biol.*, **14**, 153 (1965).

		Co-transductants / Total transductants	Mean co-transduction frequency
A. Selected marker: $supC^+$			
(i)	$lacZ2\ supC$ (x) $supC^+\ pyrF$	250/1910	0.13
(ii)	$lacZ2\ supC$ (x) $supC^+\ trpA$	150/603	0.25
(iii)	$lacZ2\ supC$ (x) $supC^+\ trpC$	997/2097	0.48
(iv)	$lacZ2\ supC$ (x) $supC^+\ ton^r$	119/161	0.74
B. Selected marker: $pyrF^+$			
(v)	$pyrF$ (x) $trpA$	539/898	0.60
(vi)	$pyrF$ (x) $trpC$	1047/1907	0.55
(vii)	$pyrF$ (x) ton^r	154/321	0.48
(viii)	$lacZ2\ pyrF$ (x) $supC^+$	357/2802	0.13

supC . . . trpC . . . pyrF. Similar reasoning with respect to *trpA* and *ton^r* establishes the complete sequence

$$supC \ldots ton^r \ldots trpC \ldots trpA \ldots pyrF$$

The co-transduction method is, as in the preceding example, most useful for establishing the relative order of three or more genes. In theory the method can also be used to determine the order of two mutant sites within a gene in relation to an outside marker; for example, if *trpC1* and *trpC2* are different mutation within *trpC* we might wish to determine whether the order is *trpC1 . . . trpC2 . . . pyrF* or *trpC2 . . . trpC1 . . . pyrF*. However, the frequencies of *pyrF–trpC1* and *pyrF–trpC2* co-transduction are unlikely to differ by more than 1–2 per cent and unrealistically large numbers of transductants would have to be analysed to obtain co-transduction frequencies that are significantly different. In this situation the three-point transduction is a more practicable and more reliable method.

2.3.2   The three-point transduction method

This is a far superior method for establishing linkage order and, as with the three-point test cross in higher organisms, it seeks to order the markers so as to account for the data with the minimum number of cross-overs — thus the larger classes of transductants are the result of two cross-overs (the minimum) while the least frequent classes are the most likely to be due to multiple crossing-over. This method is particularly useful in ordering very closely linked sites in relation to a less closely linked outside marker; in these circumstances the co-transduction method fails as it is not practical to obtain co-transduction frequencies that are statistically different.

Consider some further data on three-point transduction collected by Ethan Signer, using three of the markers already mapped by the co-transduction method (table 2.2). There are only three possible gene orders and, first, let us assume this is *trpC–supC–pyrF* (figure 2.5, upper box). With this order, whether the selected marker is *trpC^+* or *pyrF^+*, the *supC trpC^+ pyrF^+* transductants can only be explained if there are four cross-overs while the other three classes require only two cross-overs. In both selections this was the largest class of transductant and so it is very unlikely that the order is *trpC–supC–pyrF*. Similar reasoning excludes the order *supC–pyrF–trpC*.

Now consider the remaining order, *supC–trpC–pyrF*. When selection is made for *trpC^+* transductants (figure 2.5, lower left) all four classes of *trpC^+* transductants can be accounted for by two cross-overs and there is no 'rare' multiple cross-over class; this is in agreement with the data. When *pyrF^+* is the selected marker the *supC^+ trpC pyrF^+* transductants can only be accounted for if there are four cross-overs (figure 2.5, lower right) and this class is expected to be 'rare'; only one such transductant was observed among a total of 584. All these deductions indicate the gene order *supC–trpC–pyrF*, confirming the results of the previous experiments.

Table 2.2
Three-point transduction data

An *E. coli* *supC trpC pyrF* recipient was transduced with P1 grown on a *supC⁺ trpC⁺ pyrF⁺* donor and either *trpC⁺* or *pyrF⁺* transductants selected. Each transductant was then tested for the inheritance of each unselected marker. The data show the numbers of transductants recovered in each class; the markers inherited from the donor are underlined.

Adapted from Signer, Beckwith and Brenner (1965)

$$supC \ trpC \ pyrF \quad (\text{x}) \quad supC^+ \ trpC^+ \ pyrF^+$$

(i) Selecting $trpC^+$ transductants		(ii) Selecting $pyrF^+$ transductants	
$supC^+ \ \underline{trpC^+} \ pyrF^+$	102	$supC^+ \ trpC^+ \ \underline{pyrF^+}$	84
$\underline{supC^+ \ trpC^+} \ pyrF$	115	$\underline{supC} \ trpC^+ \ \underline{pyrF^+}$	250
$supC \ \underline{trpC^+} \ \underline{pyrF^+}$	257	$\underline{supC^+} \ trpC \ \underline{pyrF^+}$	1
$supC \ \underline{trpC^+} \ pyrF$	119	$supC \ trpC \ \underline{pyrF^+}$	249

Figure 2.5
Establishing gene order by three-point transductions

This figure is explained in the text.

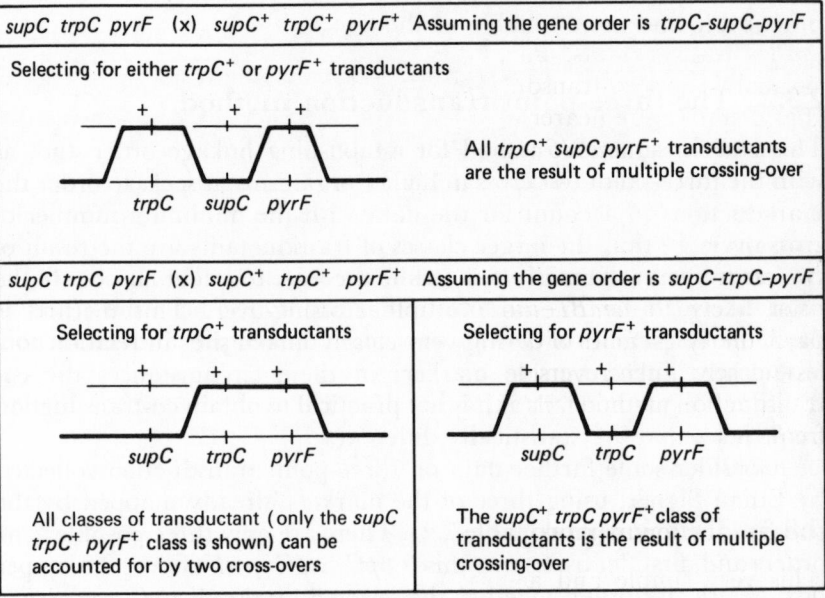

In other words, if the central marker is the one selected, then the two outsider markers will appear independently of each other among the transductants. However, if one of the outside markers is selected, then most of the transductants inheriting the other outsider marker will also inherit the central marker.

The same method can be used to order different sites within the same gene. In 1959 **Julian Gross and Ellis Englesberg** ordered 16 different mutant sites within the *ara* (arabinose) operon of *E. coli* using co-transduction, three-point and four-point transduction analyses. The data from reciprocal three-factor crosses involving three of these mutants, all located within the *araB* gene (these lack the enzyme ribulokinase and cannot use arabinose as a sole carbon source), are set out in table 2.3

Consider the first pair of reciprocal transductions, selecting for

Transduction	% ara^+ leu^+ co-transductants	Inferred gene order
leu-1 araB1 (×) *araB16*	58.1	} *araB16–araB1–leu-1*
leu-1 araB16 (×) *araB1*	28.2	
leu-1 araB16 (×) *araB23*	25.0	} *araB16–araB23–leu-1*
leu-1 araB23 (×) *araB16*	62.0	
leu-1 araB23 (×) *araB1*	32.0	} *araB23–araB1–leu-1*
leu-1 araB1 (×) *araB23*	50.8	

Table 2.3
Three-point transduction for establishing the order of sites with a gene

araB1, *araB16* and *araB23* are different mutations within the *araB* gene of *E. coli*. In each transduction *araB*⁺ transductants were selected on minimal medium supplemented with leucine and with arabinose as the sole carbon source. Each transductant was then scored for leucine requirement/ independence.

The data are adapted from Gross, J. and Englesberg, E., *Virology*, **9**, 314 (1959).

$araB^+$ transductants. Pairs of sites within the *leu* and *ara* operons normally show co-transduction frequencies between 0.55 and 0.60, so that if *araB1* were nearer to *leu-1* than *araB16* then, in the transduction *leu-1 araB1* (×) *araB16* (figure 2.6 (A)) about 55–60 per cent of the $araB^+$ transductants would also be leu^+. However, in the reciprocal transduction *leu-1 araB1* (×) *ara B16* (figure 2.6 (B)), $araB^+$ leu^+ co-transductants could only arise by multiple crossing-over and so would be less frequent. This is what was observed, so the order of the sites must be *araB16–araB1–leu-1*. Note that had *araB16* been the nearer to *leu-1*, the frequencies of co-transduction in the reciprocal transductions would have been reversed.

Similar reasoning establishes the orders *araB16–araB23–leu-1* and *araB23–araB1–leu-1* and, combining all three inferences, the order must be *araB16–araB23–araB1–leu-1*.

2.3.3   Deletion mapping

This very simple and accurate method of mapping was developed between 1959 and 1962 by **Seymour Benzer** while constructing a very fine structure genetic map of the *rIIA* and *rIIB* genes in phage T4; it is a technique that is widely used in the analysis of genetic fine structure by recombination and it is not restricted to transduction analysis. A deletion mutant is missing a part of the chromosome and, if a deletion mutant is crossed with a point mutant (that is, where mutation has affected only a single base pair), the outcome will depend on the

Figure 2.6
Three-point transduction for establishing the order of sites within a gene

If *araB1* is farther from *leu-1* than *araB16* then in the transduction *araB16 leu-1* (x) *araB1*, many of the $araB^+$ transductants will also be leu^+ (A); however, in the transduction *araB1 leu-1* (x) *araB16* there will be very few $araB^+leu^+$ transductants as they can only arise by multiple crossing-over (B).

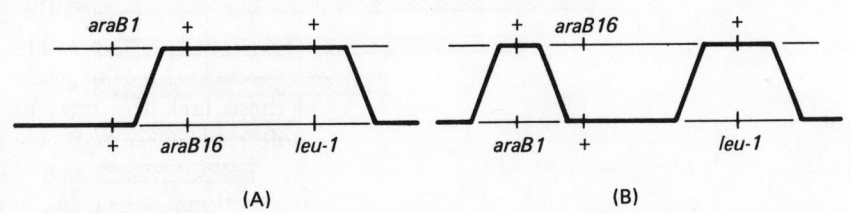

(A)   (B)

position of the point mutation in relation to the segment of DNA that is missing in the deletion mutant. If the point mutation is located within the same segment of the chromosome that is missing in the deletion mutant, it is clear that recombination cannot generate a wild type recombinant; on the other hand if the point mutation lies outside this segment, then recombination can produce a wild type recombinant.

Thus if recombinants, in this instance transductants, are observed, it means that the point mutation lies **outside** the segment covered by the deletion and by crossing a point mutant, whose map location is not known, with a series of overlapping deletion mutants, it is possible to assign the point mutation to a particular segment of the chromosome. In just the same way two different deletion mutants cannot recombine to produce a wild type recombinant if the deletions overlap as both are missing a common segment of the chromosome, whereas two non-overlapping deletions can so recombine.

The *tonB* (resistance to T1) locus of *E. coli* is particularly susceptible to deletion formation and many of the deletions originating within *tonB* extend into and terminate within the adjacent tryptophan operon. By 1971 **Charles Yanofsky and his colleagues** had isolated many different *tonB trp* deletions and used these to construct a detailed map of the tryptophan operon; some of the deletions terminating within *trpD* are shown in figure 2.7. The *trpD* gene is 1596 bp long and, by the 11 deletion mutants shown, is divided into 12 segments with an average length of 133 bp. If an unlocated *trpD⁻* point mutant is crossed with each of these deletions and wild type transductants are only recovered with AD9, AD6, AD5 and AD8, then the unmapped mutation (in this instance *trpD1081*) must lie within a segment that is common to all the other seven deletion mutants — this can only be segment VIII. Likewise, *trpD1537* only recombines with AD8 and so must be located within segment XI. The positions of *trpD1081* and *trpD1537* within segments VIII and XI can now be determined by

Figure 2.7
Overlapping deletions originating within *tonB* and terminating within *trpD*

The heavy line represents the *E. coli* chromosome and the shaded bars the segments of chromosome missing in each of the 11 deletion mutants; the zig-zag lines at the right of the boxes indicate that the deletions extend through the rest of the *trp* operon into *tonB*. The order of the genes in this region of the chromosome is

cysB . . . trpO E D C B A . . . tonB.

Adapted from Yanofsky, C., Horn, V., Bonner, M., and Stasiowski, S., *Genetics,* **69**, 409 (1971).

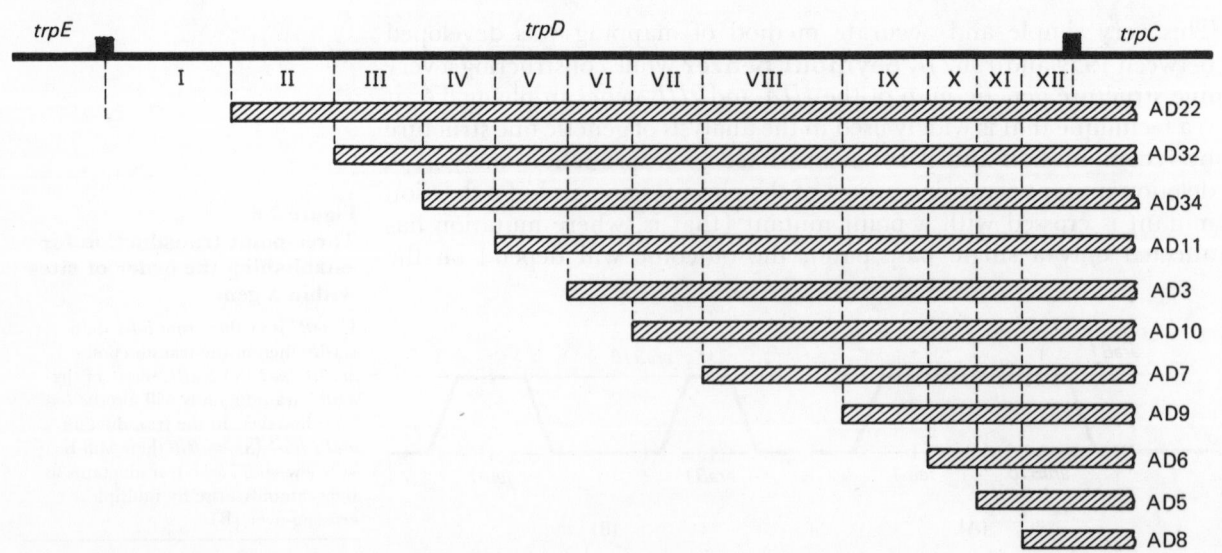

crossing with other point mutations known to be located within these segments.

With this method the labour involved in mapping is greatly reduced and, because it depends **only** on the presence or absence of wild type transductants and not on the frequencies of different classes of transductant, the accuracy of mapping is very greatly increased.

The only requirement for deletion mapping is a series of suitable deletion mutants; such mutants are available for many, but for by no means all, regions of the *E. coli* chromosome.

Deletions arise spontaneously in certain regions of the chromosome and their occurrence seems to be associated with short, directly repeated sequences—deletion removes one of these sequences and all the DNA in between them. It is thought that two different mechanisms are involved in the generation of spontaneous deletions. One mechanism involves a RecA-dependent recombination event occurring between the two repeated sequences, while the other is RecA-independent and involves strand slippage during replication (figure 2.8).

Deletions are frequently induced as a consequence of the presence of a transposable element, when a segment of DNA immediately adjacent to the element is lost. The insertion sequence ISI (section 7.5) is particularly effective in deletion formation (10^{-4} to 10^{-3} cells), while among lysogens isolated following infection with the transposable bacteriophage Mu, 10–15 per cent have lost a segment of bacterial DNA adjacent to the prophage.

2.4 Specialised transduction

In generalised transduction **any** marker on the bacterial chromosome or on a molecule of plasmid DNA can normally be transduced; this is possible because the lytic infection of *E. coli* by P1 produces rare transducing particles which, instead of packaging a molecule of phage DNA, package small random fragments of bacterial DNA. In **specialised** or **restricted** transduction, illustrated by the λ-*E. coli* system, the situation is quite different; transducing particles are only produced following **induction** of a lysogenic donor strain, the transducing particles carry a **defective** phage genome where part of the phage chromosome has been replaced by a segment of bacterial DNA, and it is **only** those genes immediately adjacent to the *attB* site, either to its left or to its right, that can be transduced. When a λ lysogen is induced most of the prophages are excised normally (section 1.5.2 and figure 1.14) but very occasionally (about 10^{-5}) an illegitimate cross-over occurs between non-homologous regions on the phage and *E. coli* chromosomes (figure 2.9). When this occurs beyond *gal* the excised phage genome has the genes from its right-hand end (the *HIJ* region) replaced by the *gal* region from the bacterial chromosome. In the same way an illegitimate cross-over beyond *bio* will produce a transducing particle carrying the *E. coli bio* region and missing the left-hand end of the phage genome. These transducing particles are known as λ_{gal} and λ_{bio} respectively.

Figure 2.8
Possible mechanisms of deletion formation

The directly repeated sequences involved in deletion formation are represented by the heavy lines. In the lower diagram only the events occurring on one limb of the replication fork are shown—the other limb is presumed to replicate normally.

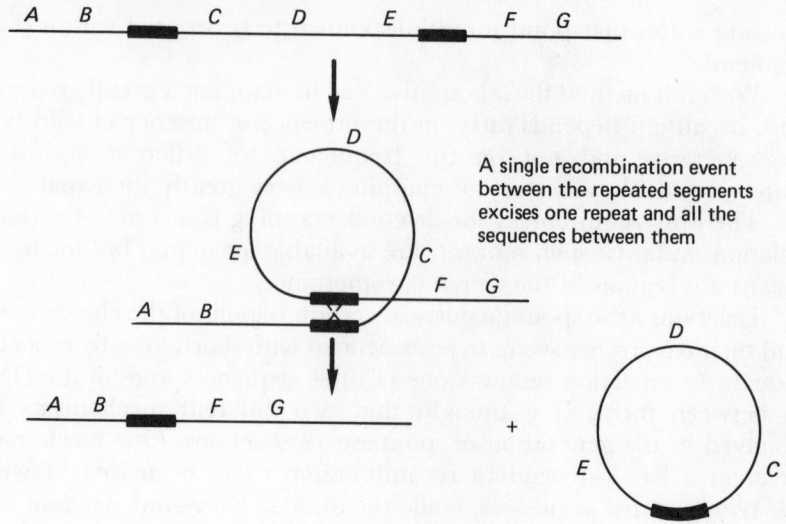

A single recombination event between the repeated segments excises one repeat and all the sequences between them

(A) Deletion by recombination

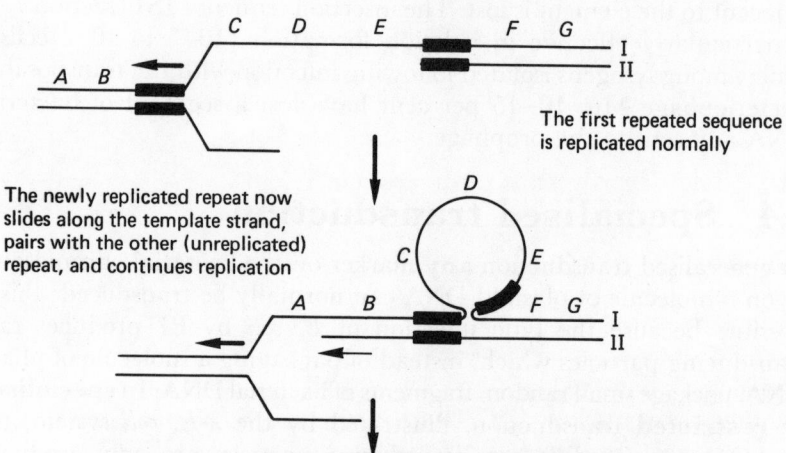

The first repeated sequence is replicated normally

The newly replicated repeat now slides along the template strand, pairs with the other (unreplicated) repeat, and continues replication

After one further replication, strand II will form a duplex molecule with one repeat sequence and all the intervening sequences deleted

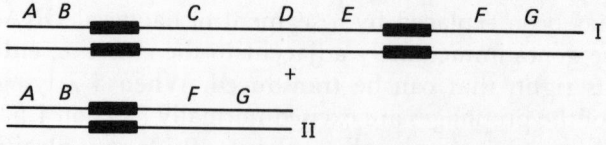

(B) Deletion by strand-slippage

Because these transducing particles are missing about 25 per cent of their DNA, including one or more essential genes, they are unable to complete the lytic cycle when they infect sensitive *E. coli* cells; this is only possible if a wild type λ^+ is also present (as in double infections), as then the essential functions that are missing in the defective λ_{gal} (or λ_{bio}) are provided by the wild type λ^+. However, the transducing particles

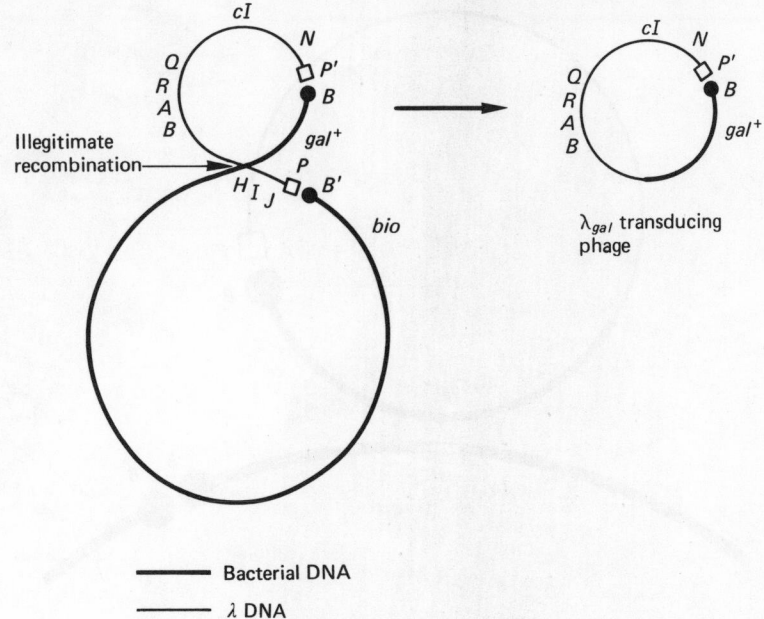

Figure 2.9
The formation of λ_{gal} transducing phage

When a λ lysogen is induced the normal excision mechanism very occasionally fails and, instead, an illegitimate cross-over occurs between non-homologous regions of the phage and bacterial chromosomes. This produces a phage missing the *HIJ* region, having instead a segment of bacterial DNA which includes the *gal*$^+$ genes.

are able to integrate into the bacterial chromosome and they confer immunity to superinfection upon the host cell.

2.4.1 Low frequency transduction

In specialised transduction, discovered in 1956 by **Melvin Morse and Joshua and Esther Lederberg**, the first step is to induce (for example) a *gal*$^+$ strain of *E. coli* lysogenic for λ. The progeny phage are harvested, used to infect a non-lysogenic *gal*$^-$ *E. coli* recipient and *gal*$^+$ transductants selected by plating on minimal medium containing galactose instead of glucose; about one transductant is recovered for every 10^5–10^6 infecting phages — this is LFT or **low frequency transduction**.

These *gal*$^+$ transductants are of two types. About one-third are stable haploid *gal*$^+$ and, as in general transduction, the *gal*$^+$ gene from the donor is recombined into the recipient chromosome replacing the *gal*$^-$ gene (figure 2.10). The remaining transductants are highly unstable and persistently produce *gal*$^-$ segregants. These transductants are partial diploids and in addition to the *gal*$^-$ gene on the recipient chromosome they have a *gal*$^+$ gene carried on the λ_{gal} genome, which by now has integrated into the recipient chromosome at *attB*; occasionally an entire λ_{gal} prophage can be excised from the recipient chromosome and, whenever this occurs, the *gal*$^-$ haploid genotype will be restored (see figure 2.12).

In low frequency transduction, because of the rarity of transducing particles, it is usual to use a high ratio of phage to bacteria, so that every *gal*$^+$ transductant is infected by not only a λ_{gal} particle but also by a normal λ^+ phage. It is thought that λ^+ first integrates at *attB* and generates *BP'* and *PB'* hybrid *att* sites, just as in normal

Figure 2.10
Stable _gal_⁺ transductants

General recombination has taken
place between the two homologous _gal_
regions with the result that the donor
gal⁺ gene has replaced the _gal_⁻ gene
in the recipient.

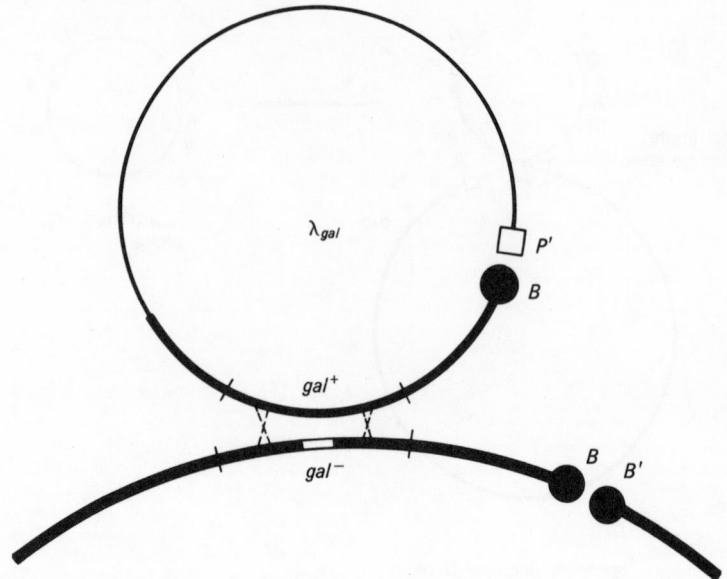

lysogenisation (figure 2.11); this is followed by the integration of λ_{gal}
at one or other of these hybrid _att_ sites by site-specific recombination.

In practice, however, nearly all the _gal_⁺ transductants have λ_{gal}
integrated into _BP'_ at the left of the λ^+ prophage. It is thought that
λ_{gal} can also integrate to the right of λ^+ by site-specific recombination
occurring between _BP'_ and _PB'_ (this is the combination of _att_ sites
involved in normal excision) and generating a λ_{gal} prophage flanked
by _PP'_ and _BB'_. However, these integrations are probably only
transitory and the integrated λ_{gal} is rapidly excised; this is largely
because _PP'_ and _BB'_, the sites involved in normal lysogenisation,
recombine very efficiently so that most integrations at _PB'_ are rapidly
followed by a further site-specific recombination event between _PP'_
and _BB'_ which excises the λ_{gal} prophage.

The partially diploid _gal_⁺ colonies produced by low frequency
transduction are very unstable and they segregate _gal_⁻ cells at about
2×10^{-3} per bacterial division. Most simply these _gal_⁻ segregants
could arise by site-specific recombination between _BP'_ and _PB'_,
excising the λ_{gal} prophage; this is the reverse of the sequence shown in
figure 2.11 but since such excision occurs infrequently it cannot account
for the instability of the transductants. This instability arises because
the λ^+ and λ_{gal} prophages lie adjacent to each other and in the same
orientation, and they effectively create a **tandem duplication.** All
tandem duplications are inherently unstable because the two homologous
sequences can pair together and participate in general recombination;
this results in the loss of one copy of the duplicated region. In the
partially diploid $\lambda^+ / \lambda_{gal}$ transductants a recombination event occurring
between the homologous λ sequences (figure 2.12) excises part of the
λ^+ and part of the λ_{gal}, leaving a _gal_⁻ chromosome carrying a
recombinant but nevertheless intact λ^+ prophage. In a similar way a

Figure 2.11
Low frequency transduction

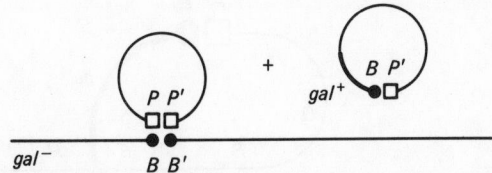

1. *gal⁻* cells are infected by both a λ_{gal^+} and by a λ^+ particle.
λ^+ first integrates at *BB'* as in normal lysogenisation

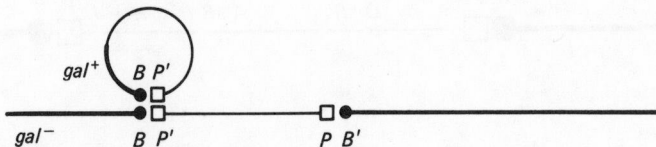

2. The *BP'* site on λ_{gal^+} can now undergo site-specific recombination with
the hybrid attachment site *BP'* (frequently) or *PB'* (rarely)

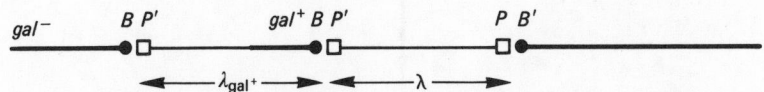

3. This integrates λ_{gal^+} into the recipient chromosome, creating a
$\lambda^+/\lambda_{gal^+}$ double lysogen

cross-over between the bacterial sequences will generate a *gal⁺* chromosome carrying a λ^+ prophage.

2.4.2 High frequency transduction

When a λ^+/λ_{gal} double lysogen is induced the defects of λ_{gal} are complemented by the homologous genes on the λ^+ prophage, and upon lysis equal numbers of λ^+ and λ_{gal} phages are released. This lysate can now be used to infect a *gal⁻* recipient at a very low multiplicity of infection and, because the lysate contains 50 per cent of λ_{gal} transducing particles there is a very high frequency of transduction (HFT or **high frequency transduction**). At such low multiplicities, most bacteria will be infected either by λ^+ or by λ_{gal} and double infections will be infrequent. In this instance the λ_{gal} genome is inserted into the recipient *gal⁻* gene by general recombination occurring between the two homologous *gal* regions (figure 2.13). This happens because the hybrid *BP'* site on λ_{gal} only recombines very inefficiently with the normal bacterial attachment site (*BB'*); λ_{gal} can only integrate at *attB* if a λ^+ integrates there first (as in low frequency transduction)

Figure 2.12
The segregants from λ^+/λ_{gal} transductants

In λ^+/λ_{gal} transductants the λ^+ and λ_{gal} prophages are arranged in tandem and they effectively constitute a tandem duplication. The bacterial chromosome can loop around so that homologous regions within λ^+ and λ_{gal} can pair together and a single recombination event can excise a λ_{gal} genome (above); this event leaves the original gal^- recipient chromosome with an intact but recombinant λ^+ prophage integrated at BB'. A similar recombination event occurring between the two gal regions (below) can generate a gal^+ lysogen.

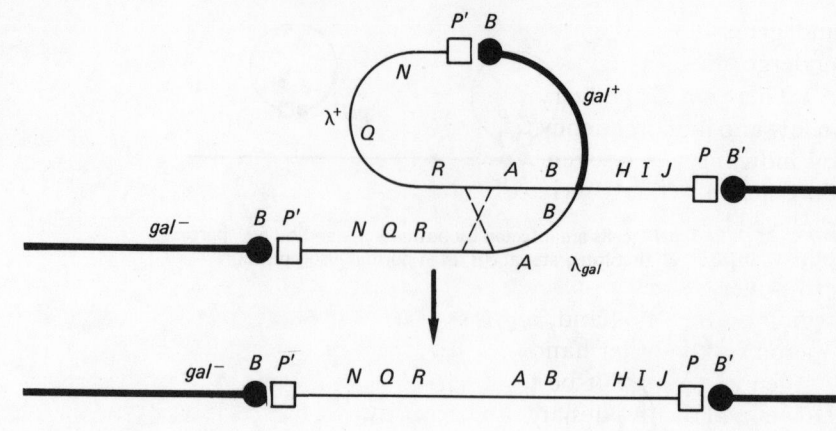

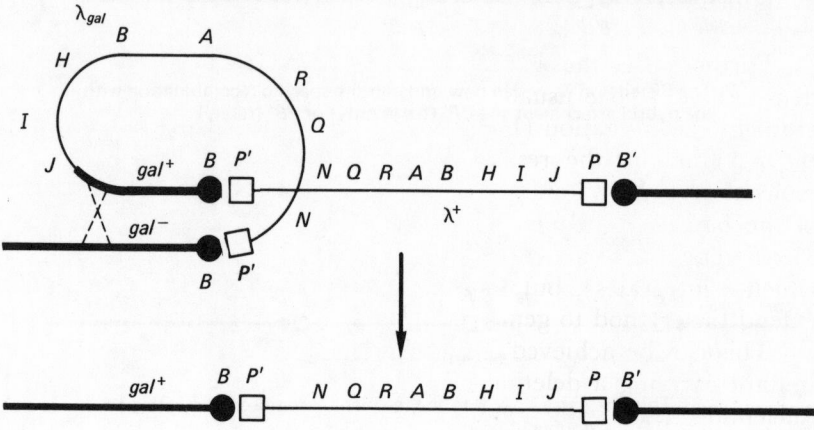

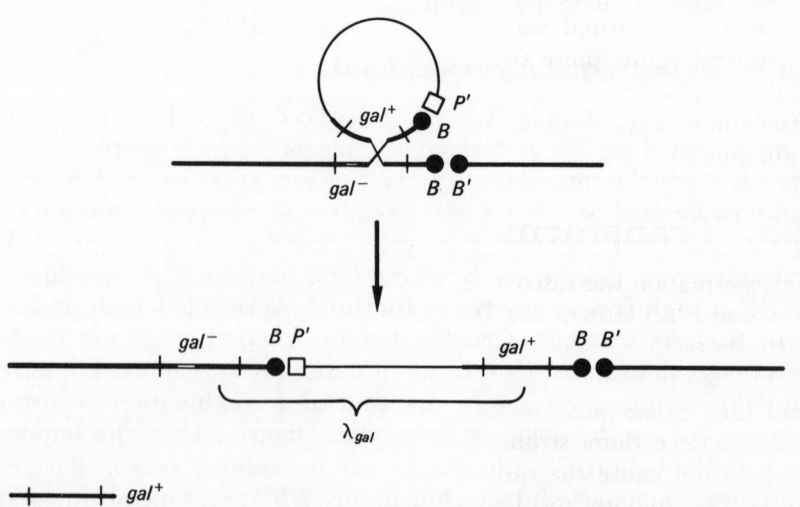

Figure 2.13
High frequency transduction

Most gal^- recipient cells are infected by a single λ_{gal^+} particle and, because λ_{gal^+} can only integrate at BB' very inefficiently, it is usually recombined into the gal region of the recipient chromosome by general recombination.

and generates new pairs of attachment sites which are then able to undergo site-specific recombination with the BP' site on λ_{gal}.

There is a further very important difference between the lysates used in low and high frequency transduction. When a LFT lysate is produced by inducing a λ lysogen it will contain a heterogeneous array of λ_{gal} particles in which varying segments of the λ chromosome have been replaced by equally variable segments of bacterial DNA. This is because many separate abnormal excisions are induced, each the result of a cross-over between a more or less random point within the phage HIJ region on the one hand, and the region to the left of the bacterial gal operon on the other hand; the principal restriction is that the excised λ_{gal} genome must not be too large to be packaged into a phage head. However, HFT lysates are made by growing and inducing a particular λ^+/λ_{gal} transductant; both the λ^+ and the λ_{gal} particles are produced by inducing **normal** excision events and, as a consequence, every λ_{gal} particle in the lysate will be identical.

Furthermore, the λ^+ and the λ_{gal} particles often have different densities and so can usually be separated from each other by density-gradient centrifugation (box 2.1). If the DNA is now extracted from the λ_{gal} fraction, the result is a homogeneous population of DNA molecules carrying a defined segment (in this instance the gal operon) of the bacterial genome. This is a very simple method for cloning *E. coli* genes very closely linked to *att* (that is, the specific BB' site at which λ integrates), but by the use of genetic tricks it is possible to extend the method to genes outside this region.

This can be achieved in two ways. Firstly, when λ lysogenises a mutant carrying a deletion of *att*λ it integrates at specific sites with nucleotide sequences resembling the sequence at *att*λ; when such a lysogen is induced, abnormal excision can produce transducing particles carrying bacterial genes closely linked to these secondary attachment sites. Secondly, chromosomal rearrangement can relocate a gene close to *att*λ and this gene can then be incorporated into a transducing particle in the usual way. In addition, the method can be further extended because there are several phages closely related to λ and each of these integrates at its own specific attachment site; for example, the attachment site for phage $\phi80$ is adjacent to the *trp* operon and induction of a $\phi80$ lysogen generates $\phi80_{trp}$ transducing particles.

2.5   Transformation

Transformation was discovered by **Fred Griffith** in 1928 in *Streptococcus pneumoniae* (pneumococcus). Pneumococci are spherical bacteria surrounded by a smooth gummy polysaccharide capsule which protects them from destruction by the normal defence mechanism of the cell, and they cause pneumonia in humans and septicaemia followed by death in mice. Some strains lack this capsule, have a rough morphology and do not cause the disease symptoms. Griffith showed that when heat-killed smooth and living avirulent rough bacteria were injected into mice, not only did many of the mice die but **living** smooth type

Box 2.1 Equilibrium density-gradient centrifugation

Density-gradient centrifugation is a technique used for separating molecules with very small differences in their density. Since (i) the density of DNA is directly related to its base composition and (ii) DNA from different sources may have different base compositions, it is frequently possible to separate two different species of DNA by this method. In the same way it is often possible to separate two populations of DNA molecules where only one is labelled with a heavy isotope (such as ^{15}N), or even two sets of phage particles containing DNA molecules with different densities (such as λ^+ and λ_{gal}).

When a strong solution of caesium chloride (or of certain other salts) is subjected to a very high centrifugal force (100 000g), the salt atoms are drawn to the base of the centrifuge tubes by the very powerful centrifugal forces; at the same time, diffusion opposes the centrifugal forces by tending to distribute the Cs^+ and Cl^- ions throughout the solution. After many hours, a state of equilibrium is reached and the balance between these opposing forces sets up a continuous concentration gradient of CsCl; the density of the solution is greatest at the base of the centrifuge tube and least at the top. Any DNA molecules dissolved in the caesium chloride solution will gradually concentrate into narrow bands where the density of the DNA molecules exactly equals the density of the CsCl at that point. Thus heavy DNA molecules will form a band at a lower position in the tube than lighter DNA molecules.

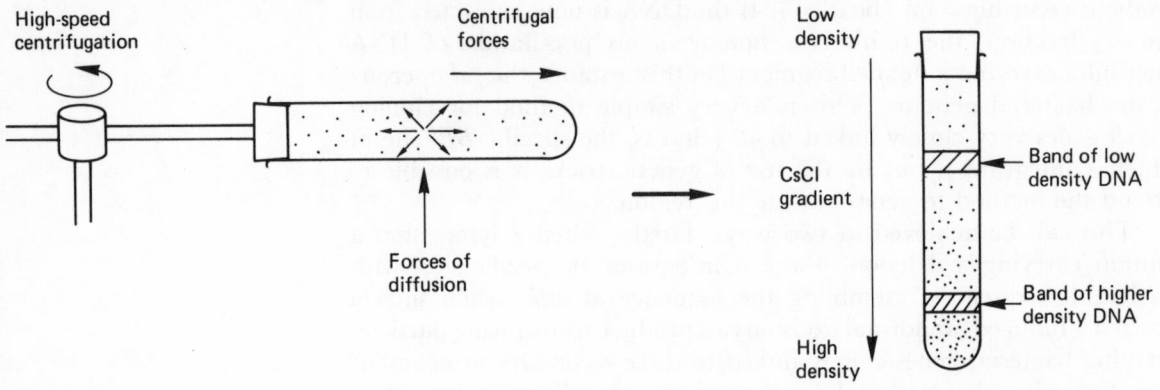

The position of the DNA in the tubes and the amounts of DNA present can be determined from the absorption of ultraviolet light. The different species of DNA (or of any other molecules being separated) can be recovered by piercing the bottom of the tube and collecting the contents drop by drop in a fraction collector.

bacteria could be recovered from the heart blood; neither the living rough nor the heat-killed smooth pneumococci caused septicaemia when injected separately. In some way the heat-killed smooth bacteria had **transformed** some of the rough bacteria to the smooth type; there appeared to be a **transforming principle** in the smooth cells which was capable of entering the rough cells, there causing a stable heritable change.

The significance of these experiments was not understood until 1944 when **Oswald Avery, Colin Macleod and Maclyn McCarty** published the results of 10 years of work with pneumococci; they had succeeded in carrying out transformation *in vitro*. They extracted the

DNA, protein and capsular components from living smooth cells and mixed each component with living rough type bacteria suspended in a synthetic medium, and they found that the DNA fraction, and only the DNA fraction, could convert some of the rough cells to the smooth type. Furthermore, the higher the purity of the DNA the more efficient was the process of transformation while, if the DNA was first treated with DNase, transformation was abolished. Hence, when DNA from cells of one genotype is incorporated into cells of a different genotype, stable heritable changes may be brought about. This was the first demonstration that DNA was endowed with specific genetic activity and was the genetic material.

Transformation, the transfer of naked DNA fragments extracted from a donor strain into a recipient strain, has been described in a variety of bacteria, including *Streptococcus pneumoniae*, *Bacillus subtilis*, *Haemophilus influenziae*, *E. coli*, *Streptomyces* spp., *Rhizobium* spp. and *Pseudomonas* spp., as well as in some blue-green algae and yeast. Transformation has been extensively used in mapping the bacterial chromosome, particularly in species such as *B. subtilis* which lack a system of conjugal transfer, and has proved a useful tool in elucidating the molecular basis of genetic recombination.

2.6   The transformation process

The following account outlines the transformation of chromosomal markers in *B. subtilis*. This involves: (1) the isolation of highly purified transforming DNA from a donor strain; (2) the uptake of this DNA by competent recipient cells; (3) the occurrence of genetic recombination between the donor DNA and the homologous region of the recipient chromosome; and (4) selection for the transformed cells (figure 2.14).

Although these stages are common to all transformation systems it must be emphasised that the details of the process may vary considerably from one bacterial species to another, particularly as to how the recipient cells develop competence and adsorb the donor DNA.

In order to carry out transformation most effectively, the transforming DNA must be double-stranded and have a minimum molecular weight of $5-6 \times 10^6$ daltons (7–9 kb of DNA, or between 1/500 and 1/400 of the bacterial chromosome) and be applied to recipient cells at a concentration of $5-10$ μg/ml. These recipient cells must be competent and in the particular physiological state which permits the uptake and integration of the transforming DNA from solution. Competent cells produce a competence factor which, in turn, probably acts by promoting the binding of the naked transforming DNA to a limited number (20–50) of receptor sites on the cell surface (figure 2.14(3)); this DNA uptake is non-specific so that any double-stranded DNA of suitable molecular weight can enter the recipient cells, but if very high molecular weight DNA is used it is cleaved by endonuclease into double-stranded fragments of $1-2 \times 10^7$ daltons (15–30 kb). As the molecules of transforming DNA cross the cell membrane, and so 'enter'

Figure 2.14
Transformation

(1), (2) DNA is isolated from a donor strain as high molecular weight fragments; if desired, these can be sheared to generate more or less uniformly sized fragments. A^+ and B^+ are closely linked and may be included on the same fragment of donor DNA. C^+ is unlinked to either A^+ or B^+ and is always on a separate fragment.

(3) A molecule of donor DNA carrying A^+B^+ binds to a receptor site of an $A^-B^-C^-$ recipient cell.

(4) The donor DNA enters the recipient cell and, as it crosses the cell membrane, one strand is degraded.

(5) The remaining single strand of A^+B^+ donor DNA is now recombined into the homologous A^-B^- region of the recipient chromosome. One strand of the duplex is displaced and the other strand pairs with the single-stranded donor DNA.

(6) All or part of the donor molecule can be integrated in this way. The remaining single-stranded segments of donor and recipient DNA are degraded (dashed lines) and the remaining single-strand gaps are ligated. The resulting heteroduplex molecule carries the nucleotide sequences corresponding to A^+ and B^+ on one strand and to A^- and B^- on the other strand.

(7) After one round of semi-conservative replication, normal base pairing is restored and one $A^+B^+C^-$ transformant and one $A^-B^-C^-$ recipient type cell is produced. Note that C^+ will only appear among the A^+B^+ transformants if two independent transformation events take place.

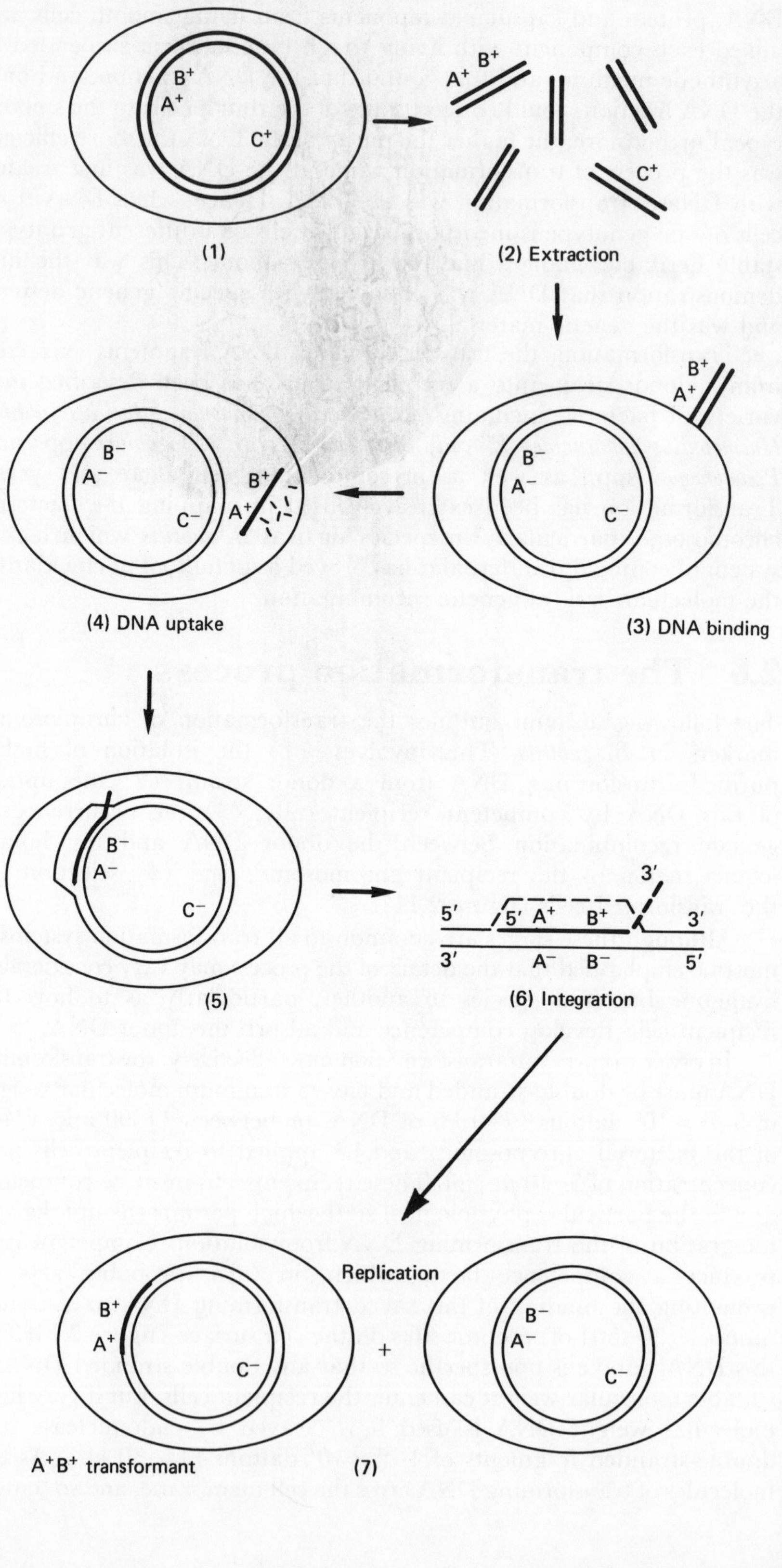

(1)

(2) Extraction

(3) DNA binding

(4) DNA uptake

(5)

(6) Integration

Replication

A^+B^+ transformant   (7)

the cell, one strand of the duplex molecule is degraded by nuclease activity (figure 2.14(4)). As a result, only one strand of donor DNA remains and this can now integrate into the homologous region of the recipient chromosome; this appears to occur by the strand of recipient DNA having the same polarity as the donor strand being displaced from the duplex molecule and replaced by the whole (or the larger part) of the fragment of single-stranded transforming DNA (figure 2.14(5) and (6)). This generates a **heteroduplex** segment within which the two strands of the duplex are of different origin and so may contain one or several differences in their nucleotide sequences. In some instances one or more of the mismatched base pairs can be corrected by **mismatch repair** or error correction (sections 5.4.1 and 6.5.2) but more usually the formation of a heteroduplex molecule is followed by a round of semi-conservative replication which produces one wild type (the A^+B^+ transformant in figure 2.14(7)) and one mutant (A^-B^-) daughter cell.

The recombination event that integrates the transforming DNA corresponds to the strand assimilation stage of the Meselson and Radding model for recombination (figure 5.4) and, like it, requires the recipient cell to have a functional system for general recombination.

It is not known why the transforming DNA has a minimum length requirement but even when much longer molecules are used it is still only possible to detect 7–9 kb of single-stranded DNA within the transformed cell, and this corresponds very closely to the length of heteroduplex DNA that is formed.

2.7 Mapping by transformation

The simplest way of mapping by transformation corresponds to the co-transduction method. The closer two markers are on the same chromosome (A^+ and B^+ in figure 2.14) the more frequently they will be included on the same fragment of transforming DNA and so the greater will be the likelihood of their being recombined into a recipient DNA molecule to form a co-transformant; hence, the shorter the map distance between the two markers the higher will be the frequency of co-transformation. Further, since transformation is a rare event (10^{-3} to 10^{-5} is the usual range) the chance of a recipient cell being simultaneously transformed by two separate molecules of donor DNA is very small (10^{-6} to 10^{-10}) and this can only occur with high concentrations of transforming DNA. Hence the close linkage can be confirmed by using the donor DNA at a lower concentration; this greatly reduces the likelihood of two separate transformation events without affecting the proportion of co-transformants.

Transformation can also provide some information on the physical distance between closely linked markers. This is because the experimenter has some control over the average size of the molecules of donor DNA; for example, if A^+ and B^+ show a high frequency of co-transformation when the average size of the donor DNA molecules is 7×10^6 daltons (10.5 kb) but a much lower frequency when the molecular weight is reduced to 5×10^6 daltons (7.5 kb) then the mutant sites within the *A* and *B* genes are probably separated by less than 3 kb of DNA.

In *B. subtilis* and *S. pneumoniae*, transformation is a well-characterised process. In *E. coli* the transformation of chromosomal markers has only been possible since 1973 and the process has been less intensively studied. However, there are at least two important differences. Firstly, in *E. coli* the cells are not normally permeable to molecules of DNA so that competence has to be induced artificially by treating the cells with calcium chloride. Secondly, *E. coli* cells produce nucleases which degrade linear molecules of DNA so that chromosomal transformation does not occur with wild type recipients; this can be overcome by using a *recBC⁻ sbcB⁻* double mutant which lacks both exonuclease V (the *recBCD* gene product) and exonuclease I (the *sbcB⁺* gene product) but which retains the capacity for general recombination by using the alternative RecE and RecF recombinational pathways (these mutations and pathways are described in section 5.8). Although the frequency of transformation is rather low (10^{-6}), a wide variety of chromosomal markers has been transformed using this method.

2.8   Plasmid transformation

Plasmid transformation is the introduction into a bacterial cell, by transformation, of a molecule of naked plasmid DNA. This technique is particularly important in genetic manipulation experiments as it provides a way to introduce almost any piece of DNA, either naturally occurring or synthesised in the test tube, into any one of a wide range of bacteria. For example, it is frequently desired to clone a 'foreign' gene into *E. coli*, the most commonly used host, where it can be replicated (that is, **cloned**) and, if required, expressed.

Firstly, recombinant DNA technology is used to insert a piece of DNA, up to 5–10 kb long and carrying the foreign gene, into a small plasmid such as pBR322 (section 3.5 and figure 3.10). This plasmid carries two genes conferring resistance to the antibiotics tetracycline and ampicillin and if, for example, the foreign DNA is inserted into a site within the gene for ampicillin resistance then the gene will be inactivated and the plasmid will only be able to confer resistance to tetracycline; this **insertional inactivation** makes it easier to detect the particular plasmid molecules carrying the insert of foreign DNA. Secondly, the recombinant plasmid is transformed into a plasmid-free *E. coli* host; since this strain does not harbour pBR322, it is sensitive to both ampicillin and tetracycline, and tetracycline-resistant clones are selected by plating on medium containing tetracycline. On this medium all the survivors will have been transformed by plasmid DNA; those carrying the insert of foreign DNA are tetracycline resistant and ampicillin sensitive and so can easily be distinguished from the tetracycline-resistant and ampicillin-resistant clones transformed by non-recombinant plasmids.

In *E. coli*, plasmid transformation involves:

(i)   treating the recipient cells with 0.1M $CaCl_2$ at $0°C$ so that they become competent;

(ii)   adding the transforming plasmid DNA which can now bind to receptor sites on the cell surface;

(iii)   heating the mixture to $42°C$ for 5 minutes, allowing the transforming DNA to enter the recipient cells;

(iv)   incubating in growth medium at $37°C$ so as to allow expression of the plasmid-encoded genes; and

(v)   recovering transformants by plating on selective medium.

This is not a very efficient process, and using supercoiled covalently closed circular DNA, about 1 transformant is recovered for every $10^4–10^5$ molecules of plasmid DNA; linear DNA molecules are transformed even less efficiently and the frequency of transformation is as low as 1 transformant per $10^8–10^9$ molecules of transforming DNA.

Plasmid transformation does not involve a recombination event so that it is not necessary for the recipient cells to have a functional recombinational system, nor is it necessary for them to be exonuclease deficient since exonucleases I and V do not attack circular molecules of DNA. However, since *E. coli* possesses a **restriction** system which recognises and attacks any foreign DNA (see box 6.1), it is customary to use a mutant which lacks a functional restriction system ($hsdR^-$ or $hsdS^-$). If this is not done the frequency of plasmid transformation may be reduced by several orders of magnitude.

Exercises

2.1   Distinguish between transduction and transformation.

2.2   The fragments of *E. coli* DNA carried by the transducing particles of P1 are of constant length but random composition, whereas those of the *Salmonella* phage P22 are of constant length *and* constant composition. How can you explain this and what effect could it have on the interpretation of transduction data involving closely linked markers?

2.3   Explain why only very closely linked pairs of markers can be co-transduced.

2.4   What may happen to a fragment of donor DNA introduced into a recipient bacterial cell by transduction?

2.5   How can transduction best be used to order a number of mutant sites within the same gene?

2.6   Critically distinguish between general and specialised transduction.

2.7   Why can only certain genes be transferred by specialised transduction? The *gal* and *bio* operons of *E. coli* cannot normally be co-transduced by λ phage. What genetic trick might enable you to co-transduce these genes?

2.8   Explain deletion mapping. Why is it such a useful method for genetic analysis?

2.9   Distinguish between low frequency and high frequency specialised transduction.

2.10 What are the requirements for carrying out transformation in *E. coli*?

2.11 Explain the importance of plasmid transformation. How does it differ from the transformation of chromosomal DNA?

2.12 In a transformation experiment, a^-b^- recipient cells were transformed either with DNA from an a^+b^+ donor *or* with a mixture of DNA extracted from a^+b^- and a^-b^+ cells. The results were as follows:

	Transformants per ml		
	a^+b^+	a^+b^-	a^-b^+
a^+b^+ donor DNA	30	1500	2000
a^+b^- & a^-b^+ DNA	25	1400	1800

What can you infer about the linkage relationships between *a* and *b*?

2.13 In *E. coli thrB* (threonine requirement), *araB* (arabinose utilisation) and *leuD* (leucine requirement) are closely linked. In the P1 transduction

$$thrB^- \; araB^- \; leuD^+ \quad (\times) \quad thrB^+ \; araB^+ \; leuD^-$$

selection was made for either $thrB^+$ or $araB^+$ transductants and each transductant characterised for the unselected marker. What can you deduce from the following results? What was the selective medium used in each instance?

	Number of transductants			
	Selecting $thrB^+$		Selecting $araB^+$	
Genotypes with respect to the unselected markers	$araB^+ leuD^+$	6	$leuD^+ thrB^+$	9
	$araB^+ leuD^-$	15	$leuD^+ thrB^-$	89
	$araB^- leuD^+$	9	$leuD^- thrB^+$	17
	$araB^- leuD^-$	270	$leuD^- thrB^-$	181

2.14 Phage P1 was grown on a wild type strain of *E. coli* and used to infect a $leu^- \; thr^- \; azi^s$ recipient; leu^+, thr^+ and $leu^+ thr^+$ were selected by plating on minimal medium + threonine, minimal medium + leucine and minimal medium. None of the thr^+ or $thr^+ leu^+$ transductants was azi^r; 60 per cent of the leu^+ transductants were azi^r and 7 per cent of the thr^+ transductants were also leu^+.

(a) Determine the order of these markers.

(b) Which two markers are the most closely linked?

(c) Why are no leu^+ thr^+ azi^r transductants found?

2.15 *lys-1, lys-2* and *lys-3* are closely linked mutations within the *lysC* gene of *E. coli*. In the following reciprocal transductions, lysine-independent transductants were selected on glucose minimal medium supplemented with methionine and the percentage inheriting the met^+ and mal^+ unselected markers determined by testing samples of the transductants. The results were as follows:

	% Lys^+ transductants inheriting	
Transduction	met^+	mal^+
met^- mal^- lys-1 (×) lys-2	32.4	7.4
met^- mal^- lys-2 (×) lys-1	1.8	81.6

Determine the order of the mutant sites. What further transductions would you carry out to determine the map position of the *lys-3* mutant site?

References and related reading

Hayes W., *The Genetics of Bacteria and their Viruses,* 2nd edn, Blackwell, Oxford (1968).

Low, K. B. and Porter, D.D., 'Modes of gene transfer and recombination in bacteria', *Ann. Rev. Genetics,* **12**, 249 (1978).

Masters, M., 'Generalised transduction', in *Bacterial Genetics* (eds J. Scaife, D. Leach and A. Galizzi), Academic Press, London, p. 197 (1985).

Notani, N. K. and Setlow, J. K., 'Mechanisms of bacterial transformation and transfection', *Prog. Nucl. Acid. Res. Molec. Biol.,* **14**, 39 (1974).

Ozeki, H. and Ikeda, H., 'Transduction mechanisms', *Ann. Rev. Genetics,* **2**, 245 (1968).

Smith, H. O., Danner, D. B. and Deich, R. A., 'Genetic transformation', *Ann. Rev. Biochem.,* **50**, 41 (1981).

3 PLASMIDS AND CONJUGATION

3.1 Introduction

It is now nearly 30 years since the F plasmid was discovered and since then many hundreds of plasmids have been found in a wide range of bacteria (and in many eucaryotes) and they are responsible for a wide range of phenotypes. Most bacterial plasmids are stably inherited and consist of an autonomously replicating circular molecule of double-stranded DNA (section 1.4); they are normally quite separate from the bacterial chromosome and, except in certain environments, they are not necessary for the survival of the host cell. They have proven to be invaluable tools in molecular biology and, for example, have been extensively used for studying the properties of bacterial genes and as cloning vectors for a variety of bacterial and eucaryotic genes.

We will consider the four most important types of plasmid:

(i)   The **F plasmid** was the first plasmid to be discovered and it has been very extensively studied. It is particularly important because it can promote the transfer of segments of the bacterial chromosome from one strain to another and this has made possible the construction of a very detailed linkage map of the *E. coli* chromosome (section 4.5).

(ii) The **R plasmids** encode resistance to one or more of a wide range of antibiotics and are very important in both medical and veterinary practice.

(iii) The **Col plasmids** encode colicins, specific bacteriocidal proteins excreted by strains of *E. coli* harbouring the corresponding Col plasmid. These strains are resistant to the colicins they produce because the plasmid also encodes a specific immunity protein but the colicin molecules will kill other strains not harbouring the Col plasmid (Col⁻) by interfering with one or other of the essential metabolic processes.

(iv)   The **recombinant plasmids** are hybrids made by re-assembling fragments derived from two or more naturally occurring plasmids and they are extensively used as cloning vectors.

Other plasmids, not considered here, may encode resistance to heavy metal ions or produce enterotoxins. The latter are of medical importance as the enterotoxins they produce are intestinal irritants responsible for travellers' diarrhoea and some types of dysentery. This only occurs if *E. coli* colonises the small intestine (it is normally found only in the large intestine) but many of the enterotoxin-producing plasmids also encode pili which assist *E. coli* to adhere to the wall of the small intestine and to effect colonisation. These plasmids also cause diarrhoea in animals.

3.2   The F plasmid

There are many different conjugative plasmids but only the conjugation system encoded by F and the related F-like plasmids have been extensively studied. The special features of F are

(i)   It is a **conjugative** plasmid and cells harbouring F (F⁺ cells) can make effective contact and form conjugal pairs with other cells not harbouring F (F⁻ cells).

(ii)   F can **mobilise** its DNA and transfer it from an F⁺ to an F⁻ cell; it can also mobilise for transfer the DNA of some non-conjugal plasmids.

(iii)   The F plasmid can recombine into the bacterial chromosome forming a **High Frequency Recombination (Hfr)** strain. In Hfr strains the bacterial chromosome behaves as if it were part of F and chromosomal genes can be very efficiently transferred from Hfr to F⁻ strains (section 4.2).

(iv)   Unlike R and Col plasmids, F does not encode any obvious phenotype.

3.2.1   The structure of F

F is a circular molecule of DNA, 94.5 kb or 30 μm in circumference (about 2.5 per cent of the size of the *E. coli* chromosome) and it is assembled from three groups of genes (figure 3.1):

(i)   A group of several genes involved in the control of vegetative replication; these have already been described (section 1.4.1).

(ii)   A group of about 28 coordinately regulated genes known as the transfer or *tra* region. These account for about one-third of the plasmid DNA and are essential for conjugation and the transfer of DNA from cell to cell.

        The transfer genes are transcribed from four separate promoters in the *traJ* region (see figure 3.1). Most of the genes

Figure 3.1
The F plasmid

F is assembled from 3 components; the transfer region which encodes the F pilus and transfer functions, the insertion sequence region and the replication region.

The transfer functions are only expressed if a functional *traJ* gene is present (lower left) – this encodes a protein which activates the transcription of both *traM* and the *tra* operon. Note that there are 4 promoters in this region and that *finP* and *traJ* are transcribed in opposite directions and that their transcripts overlap at their 5' ends.

The principal F genes and their functions are:

oriV the origin for vegetative replication.

inc, rep promote replication from *oriV*.

oriT the origin for transfer replication.

tra ALEKBVCWUNFHG required for pilus formation; *traA* encodes F-pilin, the major protein sub-unit of the pilus.

traN, traG stabilise the mating pairs of bacteria

tra MYDIZ involved in DNA metabolism during conjugation. *traI* encodes a helicase and *traY* and *traZ* act together and promote strand nicking at *oriT*.

traS, traT act to prevent mating between pairs of F⁺ bacteria.

traJ encodes a protein required to activate transcription of *traM* and the transfer operon.

finP Fertility inhibition. A regulatory gene; encodes an RNA which, when associated with the *finO* gene product, prevents expression of *traJ*. However, the *finO* gene of F is inactive as the IS3 element is inserted into it.

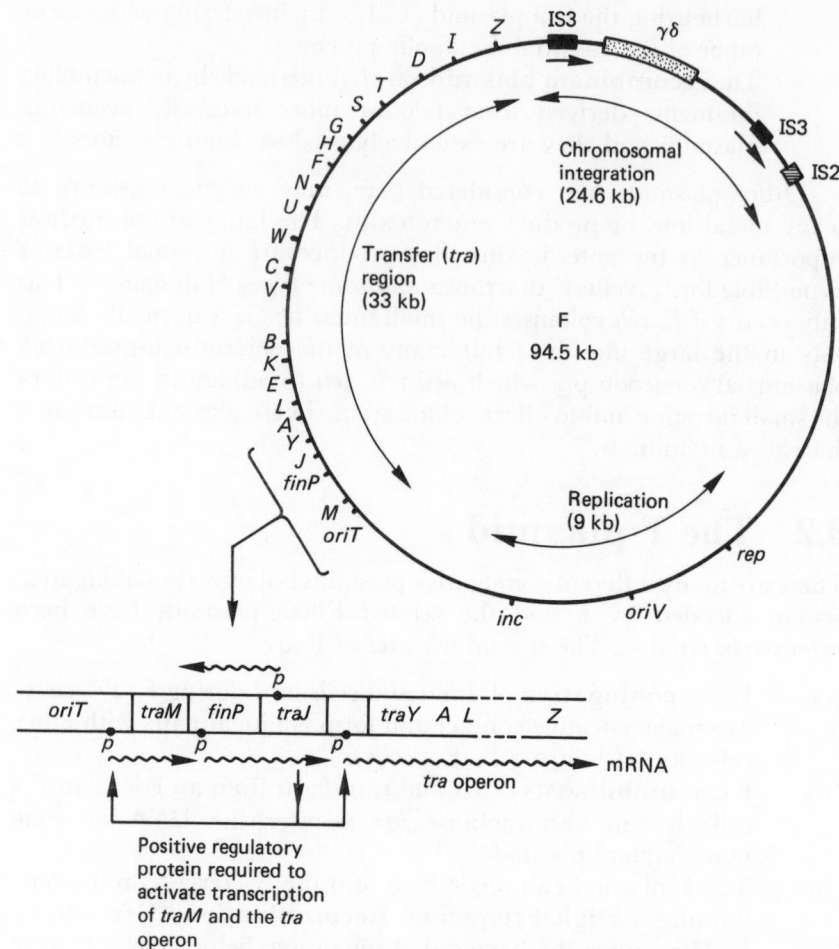

form a single very large operon about 32 kb long, the largest operon known, and this *tra* operon is only transcribed if the gene product of *traJ* is present; this gene product is also required to activate the transcription of *traM*, which lies outside the *tra* operon. Thus *traJ* positively controls the expression of all the other *tra* genes. Another control gene, *finP*, has a negative regulatory function and its role is discussed in section 3.3.1.

(iii)   A group of four transposable insertion sequences responsible for many of the unusual features of F.

This leaves a large part of the F plasmid with no known function.

3.2.2   The F-pilus

The most striking feature encoded by F is the **F-pilus** or sex-pilus, first identified by **Charles Brinton** in 1964. Many enterobacteria have flagellae, pili and other surface appendages but the F-pilus can be distinguished because it provides the specific sites to which the male-specific RNA phages R17 and MS2 adsorb. Examination of MS2 or R17 infected cells reveals that these phages form a dense coating to the F-pili, visible as long filamentous projections from the cell surface. These pili are about 8 nm in diameter but can be as long as 20 μm (typically they are about 2 μm) and they have an axial hole 2 nm in diameter. There is usually only one F-pilus per F^+ or Hfr cell and at least 14 genes in the *tra* operon are necessary for its formation.

There is no doubt that one of the functions of the F-pilus is to join together mating pairs of bacteria (F^+ and F^- or Hfr and F^-) as the ability of an F^+ cell to transfer F is strictly correlated with the presence of an F-pilus. The first stage in conjugation is the formation of a stable connection between the F^+ donor and an F^- recipient cell and it is probable that the tip of the F-pilus binds to a receptor site on the F^- cell surface and that the pilus then retracts, bringing the cells into closer contact. However, when the donor and recipient cells are allowed to aggregate, and then treated with the detergent sodium dodecyl sulphate (SDS) to cause loss of the pili, there is very little reduction in the frequency of F DNA transfer. Whereas the F-pili are absolutely essential for the formation of the closely associated pairs of donor and recipient cells, they are not obligatory for the subsequent transfer of DNA; it is probable that this close contact enables the formation of a cytoplasmic bridge, or **conjugation tube**, between the pairs of donor and recipient cells.

The F plasmid is unusual in that all the F^+ cells have an F-pilus and are able to conjugate—thus the *tra* operon is **derepressed** and is expressed constitutively. With most other conjugative plasmids, the *tra* operon is normally repressed and only switched on in about 0.1 per cent of the cells; it is only these cells that form pili and can transfer plasmid DNA.

3.2.3   The transfer of F DNA

Once the F^+ and F^- cells have formed conjugal pairs, the plasmid replicates and a copy is transferred from the F^+ to the F^- cell. This involves a special type of replication known as **transfer replication**. Whereas vegetative replication probably occurs synchronously with replication of the *E. coli* chromosome, commences at *oriV* and proceeds bidirectionally by θ-mode replication, transfer replication occurs without replication of the chromosome, commences at *oriT* (the transfer origin) and proceeds by a rolling-circle mechanism. This is similar to the rolling-circle replication of phage λ (figure 1.11) and commences by an endonuclease, thought to be encoded by *traY* and *traZ*, nicking one strand of F at *oriT*. It is always the same strand of F that is nicked and the free 5' end of this pre-existing strand is unrolled from F, probably assisted by the helicase encoded by *traI*, and passes through the conjugation tube into the F^- cell (figure 3.2) where a new complementary daughter strand is formed using the host-encoded replication machinery. This continues until a complete copy of F DNA has been transferred. Finally, an unknown mechanism circularises the linear molecule of F DNA to reform an intact F plasmid.

Figure 3.2
The transfer of F DNA

An *E. coli* chromosome is present in
each cell but has been omitted for
clarity.

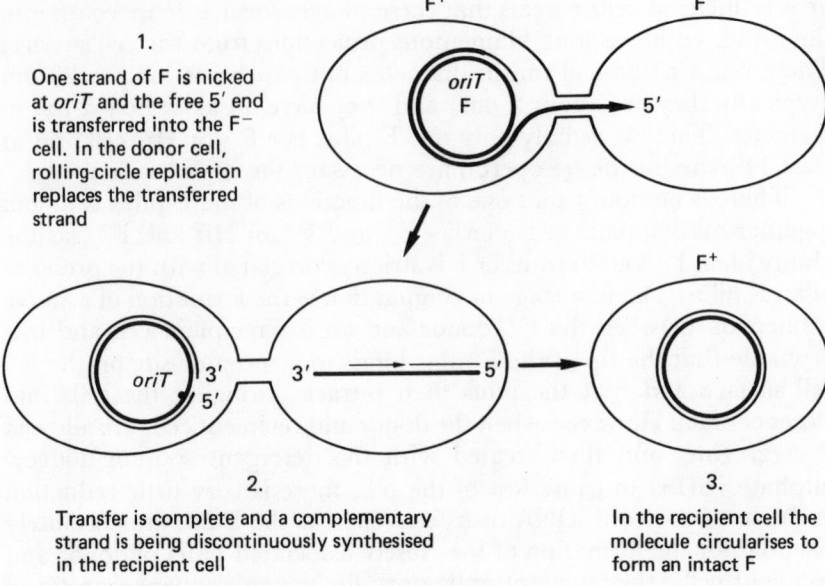

1.
One strand of F is nicked
at *oriT* and the free 5' end
is transferred into the F⁻
cell. In the donor cell,
rolling-circle replication
replaces the transferred
strand

2.
Transfer is complete and a complementary
strand is being discontinuously synthesised
in the recipient cell

3.
In the recipient cell the
molecule circularises to
form an intact F

At one time it was thought that the driving force for DNA transfer was
provided by the replication process itself but this is now thought to be
unlikely as DNA transfer can still take place when replication is blocked
in both the donor and recipient.

This conjugal transfer of F can occur twice per bacterial generation
so that an F plasmid, once introduced, can spread very rapidly through
a population of F⁻ cells.

3.2.4   The formation of Hfr strains

One of the special features of F is its ability to integrate into the bacterial
chromosome and form a variety of Hfr strains, each one able to transfer
a different segment of the bacterial chromosome into an F⁻ strain.
The introduced piece of DNA can then recombine with the F⁻
chromosome and produce genetic recombinants, a feature that has
made possible the construction of a linkage map of the *E. coli*
chromosome.

These Hfrs form because of the insertion sequences present on F.
The chromosome of *E. coli* K12 contains about five copies each of IS2
and IS3 and most Hfrs are formed by a recombination event occurring
between two identical IS elements, one on F and the other on the
E. coli chromosome (figure 3.3A); it is thought that this is promoted
by the *recA*⁺-dependent recombination machinery of the host cell.
A rather different mechanism is involved in the production of Hfrs by
the insertion sequence $\gamma\delta$, as this element is not normally present in the
E. coli chromosome. $\gamma\delta$ behave as a complex transposon and, like Tn3, it
transposes by a complex replicative mechanism (section 7.4.1); one of
the features of this process is that it can fuse together two separate
replicons (in this instance the chromosome and the F plasmid), only

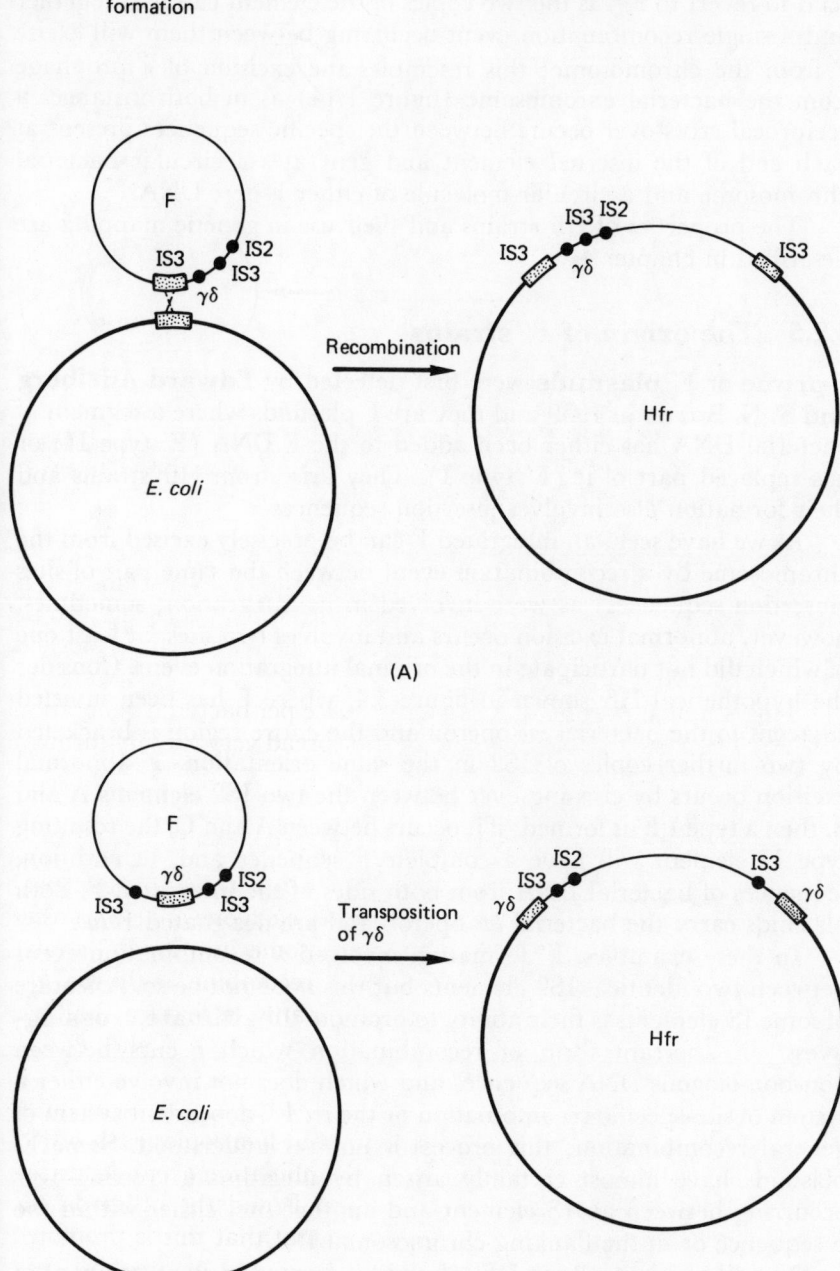

Figure 3.3
The formation of Hfr strains

Hfr strains can arise either by recombination occurring between two homologous IS elements (A) or by the replicative transposition of $\gamma\delta$ fusing together F and the chromosome (B).

one of which need carry the $\gamma\delta$ element, and at the same time generate a second copy of the element (figure 3.3B). This event only requires proteins encoded by the $\gamma\delta$ element and so can account for the occasional Hfrs that arise in F^- $recA^-$ cells.

In every Hfr, the F DNA is bracketed by a pair of directly repeated and identical IS elements. Because of this, Hfr strains are unstable and tend to revert to F^+ as the two copies of the element can pair together and a single recombination event occurring between them will excise F from the chromosome; this resembles the excision of λ prophage from the bacterial chromosome (figure 1.14) as in both instances a reciprocal cross-over occurs between the specific sequences present at each end of the inserted element and generates a circular bacterial chromosome and a circular molecule of either F or λ DNA.

The properties of Hfr strains and their use in genetic mapping are described in chapter 4.

3.2.5   The origin of F′ strains

F-prime or **F′ plasmids** were first detected by **Edward Adelberg and S. N. Burns** in 1959 and they are F plasmids where a segment of bacterial DNA has either been added to the F DNA (F′ type II) or has replaced part of it (F′ type I). They arise from Hfr strains and their formation also involves insertion sequences.

As we have seen, an integrated F can be precisely excised from the chromosome by a recombination event between the same pair of sites (insertion sequences) as were involved in its integration; sometimes, however, abnormal excision occurs and involves two sites, at least one of which did not participate in the original integration event. Consider the hypothetical Hfr shown in figure 3.4, where F has been inserted adjacent to the bacterial *lac* operon and the entire region is bracketed by two further copies of IS2 in the same orientation. If abnormal excision occurs by crossing-over between the two IS2 elements A and B, then a type I F′ is formed; if it occurs between A and C, the resulting type II element will have a complete F sequence and, in addition, sequences of bacterial DNA from both sides of the integrated F. Both plasmids carry the bacterial *lac* operon and are designated F′*lac*.

In these examples, F′ formation involved a recombination event between two identical IS2 elements but this need not be so. A feature of some IS elements is their ability to promote **illegitimate crossing-over**, an aberrant form of recombination which occurs between non-homologous DNA sequences and which does not involve either a system of site-specific recombination or the $recA^+$-dependent system of general recombination; this process is not yet understood. Some F′ plasmids have almost certainly arisen by illegitimate crossing-over occurring between an IS element and another site, either within the F sequence or in the flanking chromosomal DNA.

Note that when a type I F′ plasmid is formed, a small segment of F is left behind on the bacterial chromosome forming what is known as a **sex factor affinity locus** (*sfa*). If the type I F′ plasmid is removed (this is called **curing**) and replaced by a wild type F plasmid, then part of this F will be homologous with the *sfa* locus. As a result there is a tendency for F to integrate at this site and to form an Hfr identical to the original Hfr from which the type I F′ plasmid was derived.

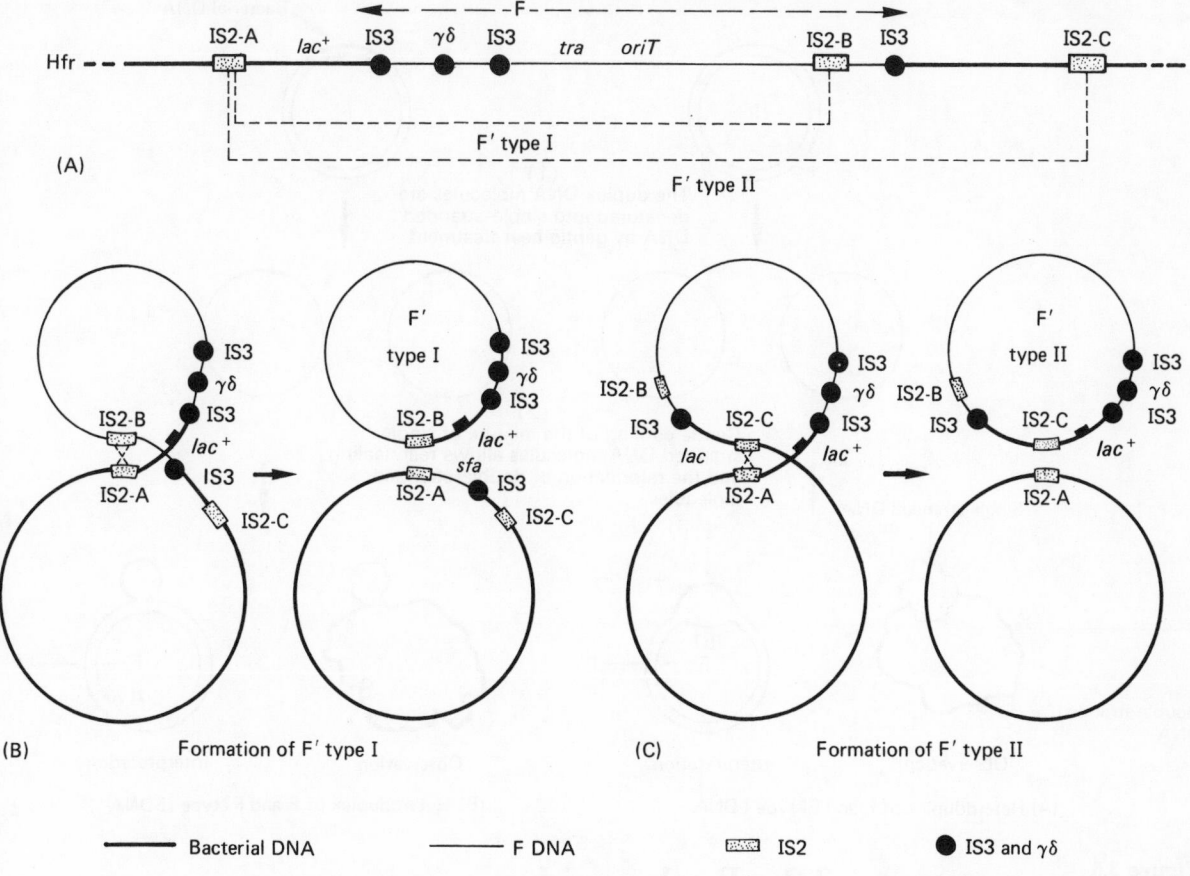

(A)

(B)   Formation of F′ type I   (C)   Formation of F′ type II

———— Bacterial DNA   ———— F DNA   ▨ IS2   ● IS3 and γδ

Figure 3.4
The formation of type I and type II F′ factors

F′ plasmids arise by recombination occurring between two homologous IS elements (A). In type I F-primes, chromosomal DNA has replaced part of the F DNA (B) but in type II F-primes, chromosomal sequences have been added to a complete F (C).

When an *sfa* locus is present in a population of F⁺ cells, up to 1 in 10 of the cells have F integrated into the chromosome at *sfa*; by way of comparison, a population of F⁺ cells without an *sfa* locus only produces random Hfrs at a frequency of 10^{-5} to 10^{-7} per generation.

A technique that has been particularly valuable in establishing the relationships between different plasmids is **heteroduplex mapping** (figure 3.5). The two plasmids to be compared are mixed together and the DNA denatured into its component single strands by gentle heating. When this mixture is cooled slowly, double-stranded molecules reform and, provided the two plasmids contain tracts of homologous nucleotide sequences, some of these will be heteroduplex, or hybrid molecules, consisting of one DNA strand from each plasmid; only homologous sequences will reform duplex DNA and any non-homologous regions will remain single-stranded. These relationships are revealed by spreading the re-annealed molecules on a grid and examining by electron microscopy. Figure 3.5 illustrates the types of heteroduplex molecule observed when F and F′ plasmids are mixed, and the interpretations of the structures.

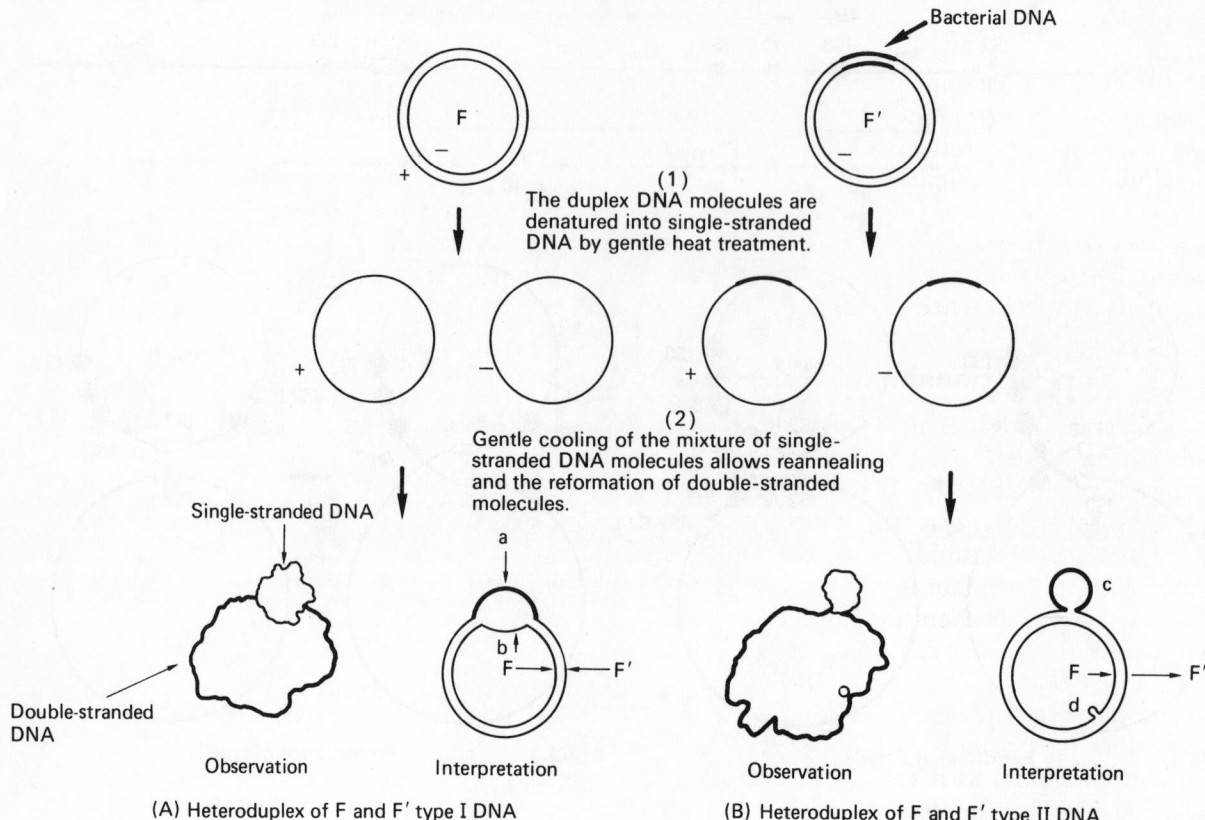

Bacterial DNA

(1)
The duplex DNA molecules are
denatured into single-stranded
DNA by gentle heat treatment.

(2)
Gentle cooling of the mixture of single-
stranded DNA molecules allows reannealing
and the reformation of double-stranded
molecules.

Single-stranded DNA

Double-stranded
DNA

Observation   Interpretation   Observation   Interpretation

(A) Heteroduplex of F and F′ type I DNA   (B) Heteroduplex of F and F′ type II DNA

**Figure 3.5
Heteroduplex mapping**

Denatured molecules of F and F′
DNA are mixed and allowed to re-
anneal. Some of the duplex molecules
formed will be heteroduplexes made
up of one strand from F and one from
F′. Regions of non-homology are seen
under the electron microscope as
tracts of single-stranded DNA.
a. Loop of bacterial DNA present on
F′ but absent from F.
b. Loop of F DNA corresponding to
the F DNA deleted during F′
formation.
c. Loop of bacterial DNA present
only on F′. There is no loop
corresponding to 'b' since all the F
sequences are present on a type II F′.
d. A small insertion (about 1 kb)
present on F.

3.3 R plasmids

R plasmids are distinguished by carrying one or more genes for drug
resistance and, because they are a threat to public health, they are of
considerable importance. When antibiotics were first introduced, a wide
variety of bacteria were sensitive to many different antibiotics and the
diseases these bacteria caused could easily be controlled. Thus *Shigella*,
the cause of bacterial dysentery in man, could be treated by the
administration of either sulphonamide or of any one of the antibiotics
streptomycin, chloramphenicol or tetramycin. However, in Japan
during the late 1950s it was first observed that *Shigella* strains isolated
from hospital patients who had been treated with just **one** of these
four common drugs had acquired multiple drug resistance to two, three
or all four of these drugs. By 1961 it was recognised that this multiple
drug resistance was infectious and could be transferred as a single unit
by cell-to-cell contact, not only among shigellae but between *Shigella*,
Haemophilus (one cause of bronchial infections and meningitis),
Escherichia and *Salmonella*, and soon after it was shown that R plasmids
were the cause of this infectious drug resistance and that they carried
one or more specific antibiotic resistance genes. One important
consequence is that *E. coli*, normally a harmless component of the
intestinal flora of both humans and livestock, can act as a reservoir for

these plasmids and transfer them to an infecting pathogen which, up to then, may not have been multiple drug resistant.

Although R plasmids appear to have been present in a small proportion of organisms even before the introduction of antibiotics, the use of antibiotics selectively favours the survival of resistant organisms and promotes dissemination of R plasmids. With the increasing and often indiscriminate use of antibiotics during the 1960s, strains of bacteria resistant to a wide range of drugs spread rapidly around the world. Thus the indiscriminate use of antibiotics, in either medicine or agriculture, is a practice that is to be discouraged at all costs.

3.3.1 The relationships between F and R plasmids

The most important R plasmids are conjugative and, like F, they are assembled from blocks of genes and, apart from the resistance determinants they carry, they may differ from F and from each other in two fundamental ways.

Firstly, two plasmids may share the same system of replication control and so regulate the number of copies per cell by identical or by very similar mechanisms. Such plasmids are **incompatible** as, if both exist in the same cell, their replication will be inhibited until the total number of copies is the same as when only one or other plasmid is present. The result is an unstable situation that is particularly important for plasmids with a very low copy number (such as F and most R Plasmids) as it very rapidly leads to the loss of one or other plasmid from all the descendent cells. As a result, two incompatible plasmids cannot normally coexist in the same cell. The resistance plasmid R386 is incompatible with F but R1, R6 and R100 all share a different system of copy number control and although they are incompatible with each other they can coexist with either F or R386.

Secondly, an R plasmid and F may have extensive regions of homology which include the transfer operons; R1, R6, R100 and R386 are all closely related to F, have nearly identical transfer operons, and are referred to as **F-like R plasmids**. Other R plasmids have different transfer operons and encode a different type of pilus.

Like F, some R plasmids are able to integrate into the bacterial chromosome but they do so at a very much lower frequency.

R100 is an F-like R plasmid but it differs from F in the way that the transfer functions are expressed. Whereas the *tra* operon of F is permanently switched on and is expressed in every F^+ (or Hfr) cell, the *tra* operon of R100 is expressed in only a small proportion (10^{-4}) of the R^+ cells and it is only these cells that produce pili and transfer the R plasmid. The *tra* operon of R100 is negatively controlled by two genes *finO* and *finP* and in the majority of R^+ cells they act to switch off the *tra* operon; *finO* and *finP* exert this control by acting together and blocking the expression of *traJ*; since the *traJ* product is a positive regulatory protein, essential for the transcription of *traM* and the *tra* operon, all the *tra* functions are coordinately switched off (section 3.2.1 and figure 3.1).

The molecular basis of this control is unusual. Until very recently it was thought that the *tra* operon was regulated by an operator–repressor mechanism, but it now appears that the active *finP* gene product is a molecule of RNA. Note that the *finP* and *traJ* genes are transcribed in opposite directions (figure 3.1) and that the transcripts overlap each other. This means that the 105 nucleotide long leader sequence of the *traJ* transcript is base complementary to the RNA transcript of *finP*; furthermore, the *finP* RNA contains an inverted repeat which can result in the formation of a stem and loop structure, and it is thought that the nucleotides in this loop interact with the base complementary nucleotides on the *traJ* leader and so prevent the expression of *traJ*. This interaction only occurs if the *finO* gene product is also present and so may require the formation of a complex between the *finP* RNA and the *finO* gene product; the latter has not yet been identified.

In contrast, the *tra* operon of F is expressed constitutively because F lacks an active *finO* gene; although F produces a *finP* RNA which binds specifically to the leader of the F *traJ* gene, this cannot prevent the expression of *traJ* because the *finO* product is absent.

Furthermore, whenever R100 is introduced into F$^+$ or Hfr cells the transfer functions of F are immediately repressed, pili are no longer present and F cannot be transferred; this is called **fertility inhibition** (Fin) and it only occurs between closely related plasmids. The most likely explanation is that the *finO* product is non-specific and can complex with the *finP* RNAs encoded by both F and R100; thus when F and R100 are present in the same cell, the R100 *finO* product forms active complexes with both the F-specific and R100-specific *finP* RNAs so that the expression of both transfer operons is blocked. This is supported by the observation that certain mutant plasmids behave as if they lack a component that is common to both the F and R100 complexes; for example, the derepressed mutant plasmid R100-1 not only expresses its *tra* functions in all cells harbouring it but it is also unable to repress the fertility functions of F.

The essential features of this model are shown in figure 3.6.

The F-like R factors have several features in common and heteroduplex mapping has shown that there is almost complete homology between the transfer segment of F and the corresponding region of R100 (figure 3.7); the only exceptions are some short sequences which are only present on R100 and others which lack sufficient homology to undergo complementary base pairing (in figure 3.7 these are the single-stranded and double-stranded loops respectively).

More important is the stem and loop structure formed within the transfer region of the R100 DNA. This structure forms because R100 has transposon Tn10 inserted at this point. The Tn10 element is about 9.3 kb long (section 7.3.1) and consists of a central sequence of unique DNA, including the gene for tetracycline resistance, flanked by a pair of insertion sequences in opposite orientations; these flanking elements form a pair of inverted repeats and the sequence of nucleotides along

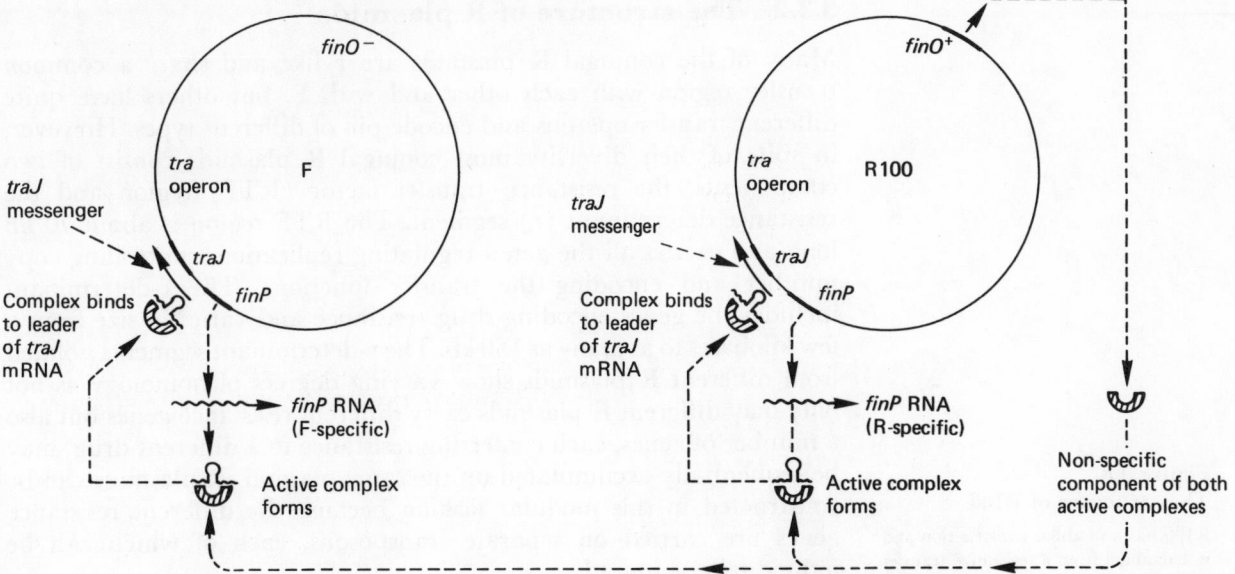

Figure 3.6
Fertility inhibition

F does not encode the non-specific
component of the *tra* operon
repressor and F is only fully repressed
when this gene product is provided by
another plasmid, such as R100.

one element is base complementary to the sequence along the other
element but in the reverse order. All these sequences are missing from
the F DNA and so when R100/F heteroduplex molecules form the
base complementary regions of the inverted repeat sequences will form
the duplex region that makes up the stem of the stem and loop structure
within the single strand of R100 DNA (reference to figure 7.3 will
make this clear).

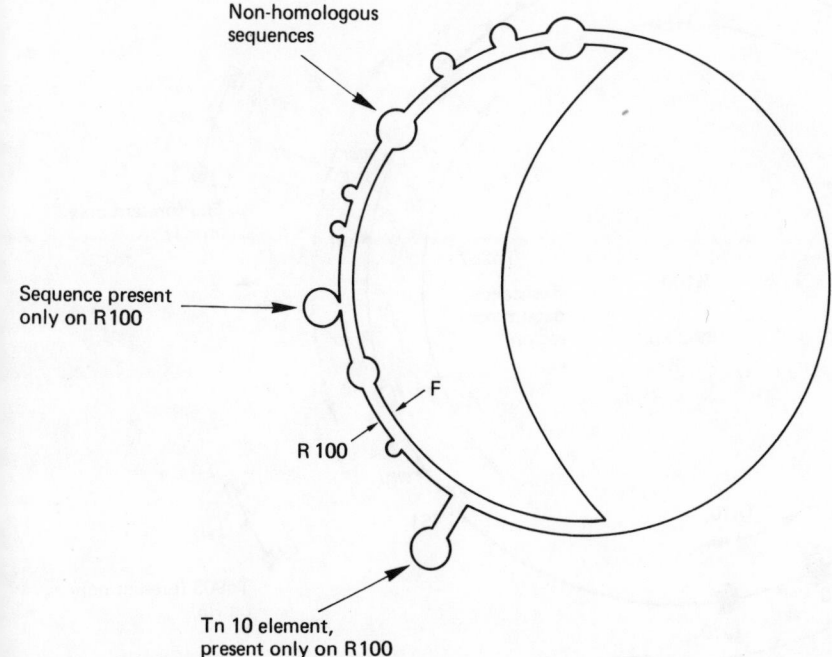

Figure 3.7
**A heteroduplex molecule of F
and R100 DNA**

F/R100 heteroduplex molecules show
that over half of R100 is homologous
to F; this region includes the transfer
operon. The region of non-homology
includes the resistance determinant
segment on R100 and the insertion
region on F. The small single and
double-stranded loops are formed by
short sequences lacking homology.
The stem and loop structure is formed
because transposon Tn10 is present on
R100 but not on F.

3.3.2   The structure of R plasmids

Many of the conjugal R plasmids are F-like and share a common transfer region with each other and with F, but others have quite different transfer operons and encode pili of different types. However, in spite of their diversity most conjugal R plasmids consist of two components, the resistance—transfer factor (RTF) region and the resistance determinant (r) segment. The RTF region is about 70 kb long and carries all the genes regulating replication, controlling copy number and encoding the transfer functions. The r-determinant includes the genes encoding drug resistance and varies in size from a few kilobases to as many as 150 kb. The r-determinant segments isolated from different R plasmids show varying degrees of homology as not only may different R plasmids carry different resistance genes but also a number of genes, each conferring resistance to a different drug, may be sequentially accumulated on the same plasmid. R plasmids can be constructed in this modular fashion because the different resistance genes are carried on separate transposons, each of which can be individually acquired.

Although R plasmids are normally stable they can, under certain circumstances, dissociate into separate RTF and r-determinant circular molecules. This is probably the result of recombination occurring between the identical IS elements flanking the r-determinant segment (see figure 3.8), and resembles the way that λ is excised from the

Figure 3.8
The structure of R100

R100 has a modular construction and is assembled from a resistance-transfer region and a resistance-determinant region. Only the tetracycline resistance gene on Tn10 is located outside the r-determinant region. The related plasmids R1 and R6 differ by having a further transposon inserted into the r-determinant region.

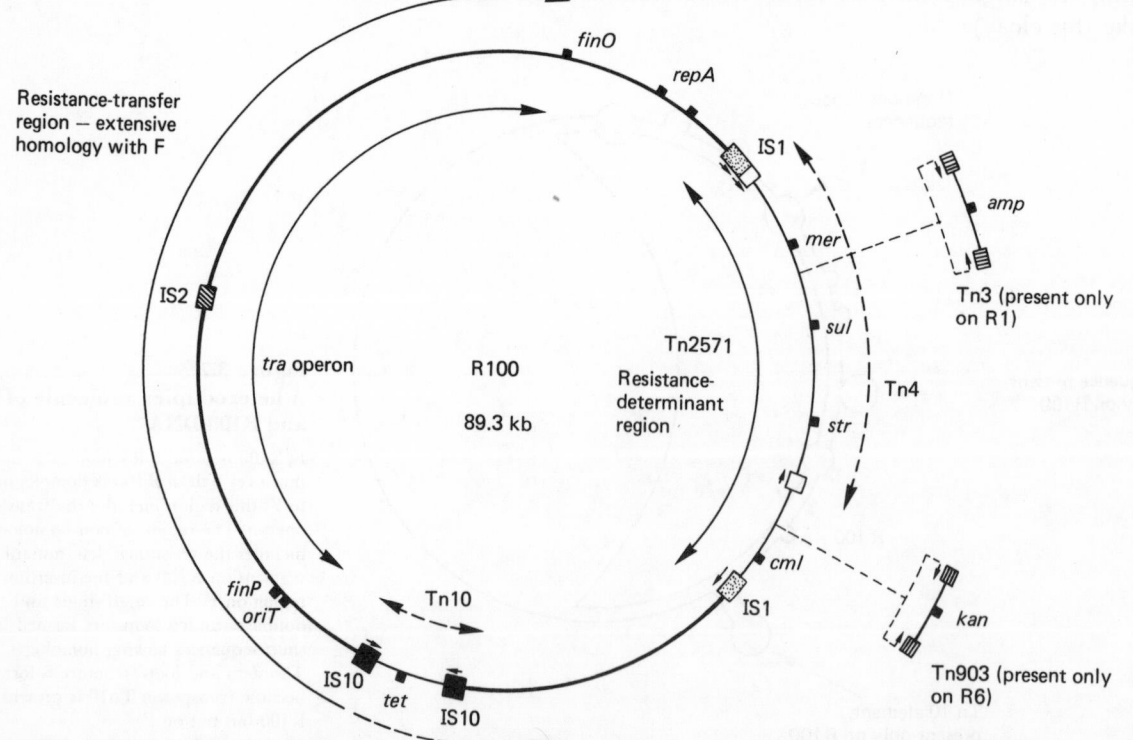

chromosome of a lysogenic cell (figure 1.14). This breakdown seems to depend on the host cell; both R1 and R100 are stable entities in *E. coli* but when either R1 is introduced into *Proteus mirabilis*, or R100 into *Salmonella typhimurium*, there is this tendency for the R plasmid to dissociate into the component RTF and r. Note that whereas RTF is a functional plasmid, the r-determinant has no replication origin and cannot exist autonomously in the cell.

The modular construction of R plasmids is seen by considering the structure of R100 and its close relatives R6 and R1 (figure 3.8), which show extensive homology with each other throughout the resistance-determinant segment. R100 is a typical R plasmid and was first isolated from *Shigella* in Japan. Over one-half the plasmid is homologous with F and this region includes the transfer operon; interestingly, a Tn10 module carrying a gene for tetracycline resistance (*tet* or *Tc*) is inserted into this region and it is the flanking IS10 modules which are responsible for the stem and loop structure shown in figure 3.7. This is the only resistance gene of R100 that is located outside the r-determinant region.

The r-determinant segment consists of a very large (23 kb) and complex transposon, Tn2571, carrying the genes for resistance to chloramphenicol (*cml* or *Cm*), streptomycin (*str* or *Sm*), sulphonamide (*sul* or *Su*) and mercury ions (*mer* or *Hg*) and flanked by two IS1 modules in the same orientation. Within Tn2571 the *str–sul–mer* genes are flanked by a pair of undesignated IS-like elements so that it is possible that Tn2571 was originally formed by a *str–sul–mer* transposon inserting into the *cml* transposon flanked by the IS1 modules. Another R-plasmid, R6, first isolated in Germany, has the almost identical structure except that a further transposon Tn903, carrying a gene for resistance to kanamycin (*kan* or *Km*), has been inserted into Tn2571. In R1, isolated in England, a Tn3 module carrying the gene for ampicillin resistance (*amp* or *Ap*) has been inserted into Tn2571 between the *sul* and *mer* genes, and this complex transposon, 15.5 kb long and flanked by the undesignated IS-like modules, is known as Tn4; in addition, there are genes for resistance to kanamycin and neomycin (*neo* or *Nm*) between IS1 and the IS-like module at the end of Tn4.

The extensive homologies between these three R plasmids, isolated from widely separated locations, would suggest that they had a common ancestor.

3.4 The Col plasmids

The colicin plasmids (section 3.1) fall into two general classes, the small non-conjugative plasmids represented by ColE1, and the much larger conjugative plasmids such as ColIb.

ColE1 is a small (6.4 kb) **multicopy** plasmid and there are normally between 10 and 30 copies per cell; this is in contrast to F which maintains a strict control over the number of plasmid copies present in a cell. The number of copies of ColE1 can be further increased to about 3000 by transferring growing cells harbouring ColE1 to medium containing chloramphenicol; this selectively inhibits the initiation of chromosome replication while permitting the continued replication of the plasmid. Because of its small size and high copy number, ColE1 and its derivatives have been extensively used as cloning vectors to provide a means for the selective amplification and isolation of particular DNA sequences that have been introduced into the plasmid.

Although ColE1 is a non-conjugative plasmid it can, like many other non-conjugative plasmids, be transferred at high frequency if a suitable conjugation system is provided by a coexisting conjugative plasmid. ColE1 can utilise the transfer functions of F or an F-like plasmid to promote its own transfer and it is said to be **mobilisable**; an F^+ ColE1$^+$ strain of *E. coli* transfers ColE1 at about the same frequency as F and over 90 per cent of the recipient cells receive both F and ColE1.

Mobilisation of non-conjugative plasmids is possible because they have both a transfer origin site (*oriT*) and one or more genes (*mob*) necessary for initiating transfer from *oriT*; the *mob* gene products substitute for the products of the F-plasmid genes *traY* and *traZ* (an endonuclease), *traI* (helicase) and *traM* (function unclear) — these are essential for initiating the transfer of F DNA but, since they act specifically at the transfer origin of F, they cannot be used by ColE1.

When ColE1 DNA is isolated it is normally supercoiled and in the form of a **relaxation complex**, and it has at least three proteins associated with it. One of these has endonuclease activity and, when activated, cuts the heavy strand of ColE1 at what is probably *oriT*, about 300 nucleotides from *oriV*; this relaxes the DNA and it assumes an open-circle configuration. One of the proteins of the relaxation complex is attached to the 5' end of the nicked strand and it has been suggested that this might act to guide this 5' end into the recipient cell. Certain mutant ColE1 plasmids are unable to form the relaxation complex and these are not mobilisable.

ColIb is a very much larger (99.1 kb) self-transmissible plasmid present in only one or two copies per cell, and although it resembles F in many of its properties it is not closely related to it. Not only are the pili encoded by ColIb of a different kind (I-type pili) and unable to adsorb the phages which attach specifically to the F-type pili, but also the two plasmids have different and non-interacting systems of replicational control; as a consequence, F and ColIb can coexist in the same cell and each produces its own type of pilus and controls its own transfer.

3.5   Recombinant plasmids

Very many recombinant plasmids suitable for use as cloning vectors have been constructed using recombinant DNA technology, and the simplest and most extensively used are derived from the small bacterial plasmids. **Cloning** is the production of a large number of identical molecules of DNA and it is possible because many plasmids (and certain phages such as M13) continue to replicate and to function normally even after a large piece of foreign DNA (up to 5–6 kb) has been inserted into them; this means that large amounts of cloned DNA can be isolated from bacteria harbouring such a **chimeric** or **recombinant plasmid**.

Although gene cloning and the techniques of recombinant DNA technology are not within the scope of this book, we will consider briefly one very important recombinant plasmid, pBR322 (p indicates plasmid and BR indicates the laboratory in which it was constructed), and show how it can be used as a cloning **vector**; pBR322 (figure 3.9) and its derivatives are among the most important cloning vectors as they have been specially constructed to have all the required essential properties. Firstly, pBR322 is only 4363 bp in size and so can be easily manipulated. Secondly, it carries two genes for antibiotic resistance, amp^R derived from the Tn3 module on plasmid R1 and tet^R from the recombinant plasmid pSC101; this is a very valuable property as host cells carrying pBR322 can be recognised by their resistance to both ampicillin and tetracycline and distinguished from other host cells carrying a chimeric pBR322 plasmid where the gene to be cloned has been inserted into (and so has inactivated) the amp^R gene — such cells will only be resistant to tetracycline (see figure 3.11). Thirdly, because the replication origin is derived from pMB1, a close relative of ColE1, pBR322 is a high copy number plasmid.

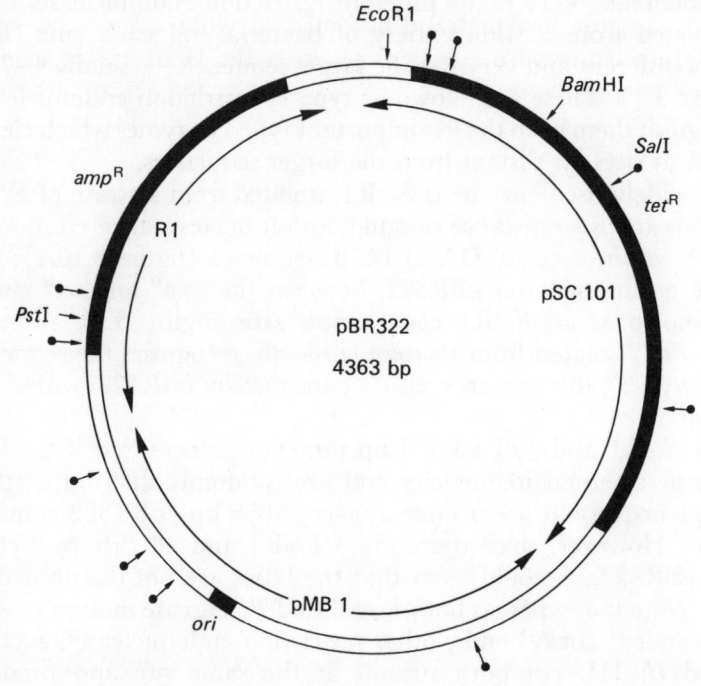

● → AluI restriction site

Figure 3.9
The recombinant plasmid pBR322

pBR322 is derived from components derived from plasmids R1, pSC101 and pMB1 (a derivative of ColE1). The figure shows the genetic map and a map of some of the restriction sites. Note that *Eco*R1, *Bam*HI, *Sal*I and *Pst*I restriction sites occur only once but that there are at least 10 *Alu*I restriction sites.

Other plasmids derived from pBR322 contain the operator — promoter regions derived from either the lactose operon or from λ, and if the gene to be cloned is inserted adjacent to these sequences then its expression can be controlled through the *lac* or λcI repressors.

At its simplest, gene cloning involves the *in vitro* insertion of a fragment of DNA carrying the gene to be cloned into a plasmid (or phage) vector, followed by the introduction of this chimeric plasmid into a host bacterial strain using the technique of plasmid transformation (section 2.8). Within the host cell the vector replicates, producing many copies of itself, and after many divisions of the host cells a clone is produced, consisting of very many bacteria each containing many identical copies of the recombinant DNA molecule. In bacteria transformed by pBR322 there are normally about 15 copies per cell, but, since the replicational origin and the system for the control of copy number are derived from ColE1, this number can be increased to 1000–3000 in the presence of chloramphenicol (see section 1.4.1). These chimeric plasmids are made using a special class of enzymes known as restriction endonucleases.

3.5.1   Restriction endonucleases

One of the most important events in the development of genetic engineering was the discovery and exploitation of **restriction endonucleases**. Very many different restriction endonucleases have been isolated from a wide variety of bacteria and each cuts DNA **within** a different and very specific target sequence — usually 4–7 bp long (box 3.1). These are known as type II restriction endonucleases to distinguish them from the less important type I enzymes which cleave the DNA at sites far distant from the target sequences.

One widely used enzyme is *Eco*R1, isolated from a strain of *E. coli* harbouring the R1 resistance plasmid, and it makes staggered nicks in the DNA within every 5′ GAATTC 3′ sequence (figure 3.10A); this sequence occurs once on pBR322, between the amp^R and tet^R genes, and is known as an *Eco*R1 **restriction site** (figure 3.9). Another enzyme, *Pst*I, isolated from *Providencia stuartii*, recognises the sequence 5′ CTGCAG 3′; this sequence also occurs once on pBR322, within the amp^R gene.

Both *Eco*R1 and *Pst*I have 6 bp target sequences and if the four bases occur at equal frequencies and are randomly distributed then each sequence would occur once in every 4096 bp (pBR322 contains 4363 bp). However, since there are 5 *Eco*R1 and 28 *Pst*I restriction sites on pBR322, it would seem that the bases are not distributed at random. Note that whereas both *Eco*R1 and *Pst*I create molecules with single-stranded 'sticky' ends, other restriction endonucleases, such as *Alu*I and *Hae*III, cut both strands at the same site and produce molecules with 'blunt' ends.

Each bacterial strain has at least one restriction endonuclease and it is these that are responsible for protecting the cell against any infecting

Box 3.1 Some important restriction endonucleases

Restriction endonuclease	Organism	Target sequence	Products		Number of target sites on λ DNA
BamHI	Bacillus amyloliquefaciens	---G↓G A T C C--- ---C C T A G↑G---	---G ---C C T A G	+	G A T C C --- G --- 5
BglII	Bacillus globigii	---A↓G A T C T--- ---T C T A G↑A---	---A ---T C T A G	+	G A T C T --- A --- 6
EcoR1	Escherichia coli R	---G↓A A T T C--- ---C T T A A↑G---	---G ---C T T A A	+	A A T T C --- G --- 5
HindIII	Haemophilus influenzae R$_d$	---A↓A G C T T--- ---T T C G A↑A---	---A ---T T C G A	+	A G C T T --- A --- 7
HpaII	Haemophilus parainfluenzae	---C↓C G G--- ---G G C↑C---	---C ---G G C	+	C G G --- >50 C ---
PstI	Providencia stuartii	---C T G C A↓G--- ---G↑A C G T C---	---C T G C A ---G	+	G --- A C G T C --- 28
SalI	Streptomyces albus	---G↓T C G A C--- ---C A G C T↑G---	---G ---C A G C T	+	T C G A C --- G --- 2
TaqI	Thermus aquaticus	---T↓C G A--- ---A G C↑T---	---T ---A G C	+	C G A --- >50 T ---
AluI	Arthrobacter luteus	---A G↓C T--- ---T C↑G A---	---A G ---T C	+	C T --- >50 G A ---
HaeIII	Haemophilus aegyptius	---G G↓C C--- ---C C↑G G---	---G G ---C C	+	C C --- >50 G G ---

Sequences are written so that the 5′ end of the upper strand is at the left. Most restriction endonucleases, like those shown here, have palindromic target sequences. Note that AluI and HaeIII generate blunt ends while all the other enzymes shown create sticky ends.

'foreign' DNA. This immediately raises the question as to why these nucleases do not attack the DNA of their own cell. The answer is that each bacterial strain not only has a specific restriction endonuclease but also a complementary **modification enzyme** or **methylase** which recognises the same target sequence and methylates particular bases within it. For example, most strains of *E. coli* contain a methylase which recognises the 5′ GAÅTTC 3′ sequence and methylates one of the adenines (Å); since these sequences are symmetrical, one adenine on each strand will be methylated, protecting the DNA from *Eco* R1 attack but not necessarily from attack by other restriction endonucleases (methylation is described in box 6.1).

The restriction endonucleases have a two-fold importance. Firstly, as we have seen, enzymes such as *Eco*R1 and *Pst*I cleave DNA into fragments by attacking specific target sequences or restriction sites,

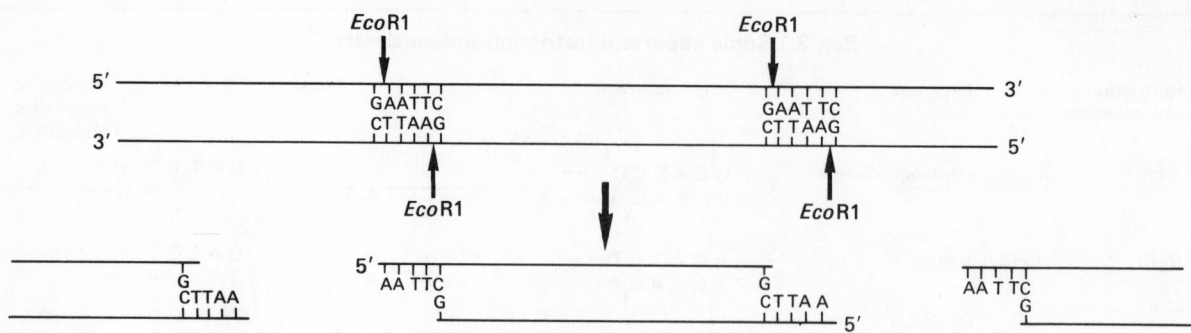

(A) *Eco* R1 cleaves DNA at every 5' GAATTC 3' sequence producing fragments of DNA with single-stranded cohesive ends

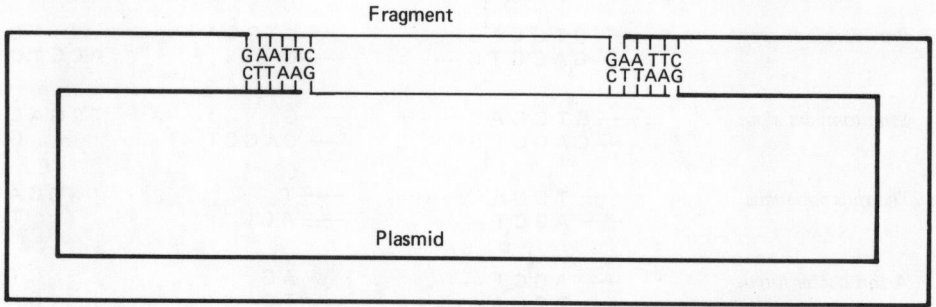

(B) If plasmid DNA with a single *Eco* R1 restriction site is also cleaved with *Eco* R1, the fragment can be inserted into the plasmid DNA and the remaining single-strand breaks sealed with ligase

**Figure 3.10
Restriction endonuclease
*Eco*R1**

creating single-stranded sticky ends at each cleavage site. Since the target sequences are palindromic all the 5' sticky ends generated by a particular restriction endonuclease will have the same base complementary sequences so that a sequence cleaved from one molecule can be annealed into a different molecule (figure 3.10B). Secondly, each restriction enzyme will convert a population of duplex molecules into a unique array of restriction fragments which can, by virtue of their different sizes, be separated by gel electrophoresis. Each restriction enzyme produces a different array of fragments and, by comparing the results from different restriction digests, it is possible to construct a **restriction map** showing the physical location of each specific restriction site on the molecule of DNA; the restriction map constructed from physical data can now be compared with the genetic map assembled from recombinational analysis.

3.5.2 Gene cloning using pBR322

At its simplest, gene cloning using a plasmid vector involves: (1) the

construction of a vector (in this example, pBR322) into which has
been inserted a fragment of DNA containing the gene to be cloned;
(2) the introduction of this recombinant plasmid into a host cell by

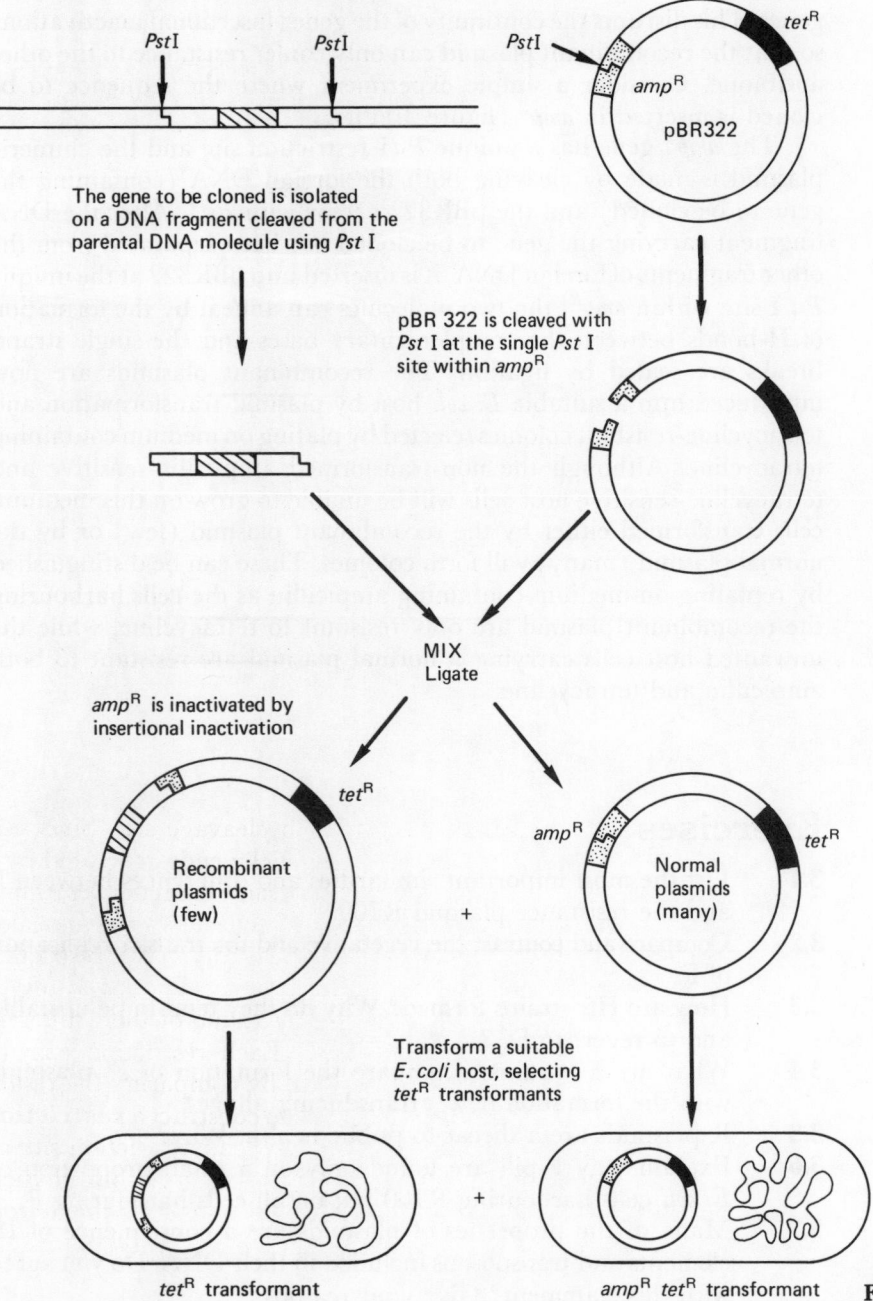

PstI   PstI   PstI   tet^R

                                                amp^R
                                                pBR322

The gene to be cloned is isolated
on a DNA fragment cleaved from the
parental DNA molecule using *Pst* I

pBR 322 is cleaved with
Pst I at the single *Pst* I
site within amp^R

MIX
Ligate

amp^R is inactivated by
insertional inactivation

tet^R

Recombinant   amp^R tet^R
plasmids
(few)   Normal
                                                                   plasmids
                                    +   (many)

Transform a suitable
E. coli host, selecting
tet^R transformants

                    +

tet^R transformant   amp^R tet^R transformant

The two classes of transformant can be distinguished by replating on medium
containing ampicillin

**Figure 3.11
A simple cloning experiment
using pBR322**

plasmid transformation (section 2.8); and (3) selection for transformed host cells containing the desired recombinant plasmid.

pBR322, in common with many cloning vectors, has two genes conferring easily recognisable phenotypes (amp^R and tet^R), and the recovery of transformed cells carrying the recombinant plasmid is facilitated if the gene to be cloned is inserted into one or other of these genes. This disrupts the continuity of the gene (insertional inactivation) so that the recombinant plasmid can only confer resistance to the other antibiotic. Consider a simple experiment where the sequence to be cloned is inserted in amp^R (figure 3.11).

The amp^R gene has a unique *Pst*I restriction site and the chimeric plasmid is made by cleaving both the foreign DNA (containing the gene to be cloned) and the pBR322 vector with *Pst*I. After the DNA fragment carrying the gene to be cloned has been separated from the other fragments of foreign DNA, it is inserted into pBR322 at the unique *Pst* I site within amp^R; the two molecules can anneal by the formation of H-bonds between the complementary bases and the single-strand breaks are sealed by ligation. The recombinant plasmids are now introduced into a suitable *E. coli* host by plasmid transformation and tetracycline-resistant colonies selected by plating on medium containing tetracycline. Although the non-transformed ampicillin-sensitive and tetracycline-sensitive host cells will be unable to grow on this medium, cells transformed either by the recombinant plasmid (few) or by the normal plasmid (many) will form colonies. These can be distinguished by replating on medium containing ampicillin as the cells harbouring the recombinant plasmid are only resistant to tetracycline, while the unwanted host cells carrying a normal plasmid are resistant to both ampicillin and tetracycline.

Exercises

3.1   List the most important similarities and differences between F and the resistance plasmid R100.

3.2   Compare and contrast the vegetative and the transfer replication of F.

3.3   How are Hfr strains formed? Why do they tend to be unstable and to revert to F^+?

3.4   What are F' strains? Compare the formation of F' plasmids with the formation of λ_{gal} transducing phages.

3.5   R plasmids are a threat to public health. Why?

3.6   Explain why F-pili are found only on a small proportion of *E. coli* cells harbouring R100 but on all cells harbouring F.

3.7   Many of the properties of plasmids are a consequence of IS elements and transposons included in their DNA. Do you agree with this statement? Give your reasons.

3.8   How would you transfer a non-transmissible plasmid, such as ColE1, from one *E. coli* strain to another?

3.9 Describe how heteroduplex mapping has contributed to our knowledge of the structure of plasmids and their interrelationships.

3.10 Compare the structure of the F and ColE1 plasmids.

3.11 Explain why pBR322 is so useful as a cloning vector.

3.12 Explain what is meant by 'plasmid incompatibility' and 'fertility inhibition'.

3.13 Explain the difference between type I and type II F′ plasmids. What is a sex factor affinity locus?

3.14 How would you clone a gene conferring resistance to kanamycin using pBR322? How would the methods differ if you inserted this gene into the intergenic region between amp^R and tet^R instead of into either amp^R or tet^R.

References and related reading

Broda, P., *Plasmids*, Freeman, San Francisco (1979).

Clark, A. J. and Warren, G. J., 'Conjugal transmission of plasmids', *Ann. Rev. Genetics*, **13**, 99 (1979).

Foster, T. J., 'Analysis of plasmids with transposons', *Methods in Microbiol.*, **17**, 197 (1984).

Hardy, K., *Bacterial Plasmids*, 2nd edn. Van Nostrand Reinhold (UK), Wokingham (1986).

Ippen-Ihler, K. A. and Minkley, E. G. 'The conjugation system of F, the fertility factor of *Escherichia coli*', *Ann. Rev. Genetics*, **20**, 593 (1986).

Willetts, N., 'Plasmids', in *Genetics of Bacteria* (eds J. Scaife, D. Leach and A. Galizzi), Academic Press, London, p. 165 (1985).

Willetts, N. and Skurray, R., 'The conjugation system of F-like plasmids', *Ann. Rev. Genetics*, **14**, 41 (1980).

4 GENETIC ANALYSIS USING Hfr AND F′ STRAINS OF *E. COLI*

4.1 Introduction

When F$^+$ and F$^-$ cells are mixed they rapidly form mating pairs and copies of F are transferred from the F$^+$ to the F$^-$ cells, converting all the latter to F$^+$. In these crosses not only is F transferred at very high frequency but also segments of the F$^+$ donor chromosome are transferred at very low frequency (10^{-6} to 10^{-7}) forming cells that are partial diploids or **merozygotes**; this transfer can be followed by recombination between the segment transferred from the F$^+$ and the corresponding region of the F$^-$ chromosome to produce rare genetic recombinants. Some researchers believe that the occasional chromosome transfer observed in F$^+$ × F$^-$ crosses is due to rare Hfr cells arising spontaneously in the F$^+$ population; however, since most of the recombinants are F$^+$ and not F$^-$ (as they are in Hfr × F$^-$ crosses) it seems unlikely that this is the only explanation.

It was the occurrence of these rare recombinants that, in 1946, enabled **Joshua Lederberg** to distinguish F$^+$ and F$^-$ strains and to construct a very simple linkage map of the *E. coli* chromosome; the only assumption he made was that the closer two genes are on the *E. coli* chromosome the more likely they are to be transferred together, and he expressed genetic distances in terms of the relative frequency of recombination between each pair of genes. Very soon afterwards **William Hayes** and **Luca Cavalli** independently isolated donor strains which produced several hundred times more recombinants than F$^+$ strains and they called these **Hfr** or *high frequency recombination* strains. It was the brilliant analyses of these Hfr strains by **Elie Wollman**, **François Jacob** and **William hayes** that led to the unravelling of the complexities of the mating process and to the development of two unique methods for mapping the *E. coli* chromosome.

4.2 Chromosome transfer by Hfr cells

In Hfr cells the F factor is inserted into the *E. coli* chromosome and, although they can no longer transfer F independently of the chromosome, the transfer functions of F are still expressed. Not only do Hfr cells have F-pili and form conjugal pairs with F⁻ cells, but this is followed by the partial and oriented transfer of the Hfr chromosome into the F⁻ recipient cells. This transfer, like the transfer of F, involves a rolling-circle type of replication, initiated by the *tra* endonuclease nicking the integrated F at *oriT*, followed by the transfer of the free 5′ end of the single strand of pre-existing Hfr DNA into the F⁻ cell (figure 4.1).

The first sequence to be transferred is the part of F which includes the insertion sequences and this is followed by the bacterial genes closest to the free 5′ end of the nicked strand; as transfer proceeds (figure 4.2) the more distal markers will be transferred in the **same sequence as they occur around the chromosome**. It is important to realise that the *tra* endonuclease **always** nicks the **same** strand of F and so a given Hfr will always transfer the same strand of Hfr DNA.

This unidirectional transfer replication takes about 100 minutes (compared with the 30–40 minutes required for bidirectional vegetative replication) and the strand of Hfr DNA is transferred at a uniform rate of about 38 kb (11 μm) per minute. However, during conjugation the cells are in a state of constant movement and shearing forces will tend to rupture the conjugation tube and separate the mating bacteria, so severing the strand of Hfr DNA during the process of transfer. These random breakages occur as a function of time and most of the conjugal pairs will separate during the first 40 minutes after mating. The consequences of these breakages are two-fold. Firstly, it is extremely rare for the entire Hfr chromosome to be transferred and only the proximal markers (that is, those nearest the origin) are transferred at

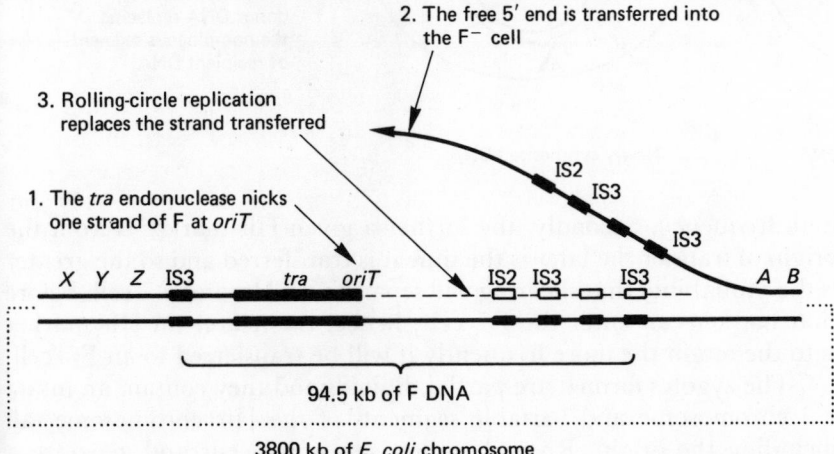

2. The free 5′ end is transferred into the F⁻ cell

3. Rolling-circle replication replaces the strand transferred

1. The *tra* endonuclease nicks one strand of F at *oriT*

X Y Z IS3   *tra*   *oriT*   IS2 IS3   IS3   A B

IS2
IS3
IS3

94.5 kb of F DNA

3800 kb of *E. coli* chromosome

Figure 4.1
The initiation of transfer of an Hfr chromosome

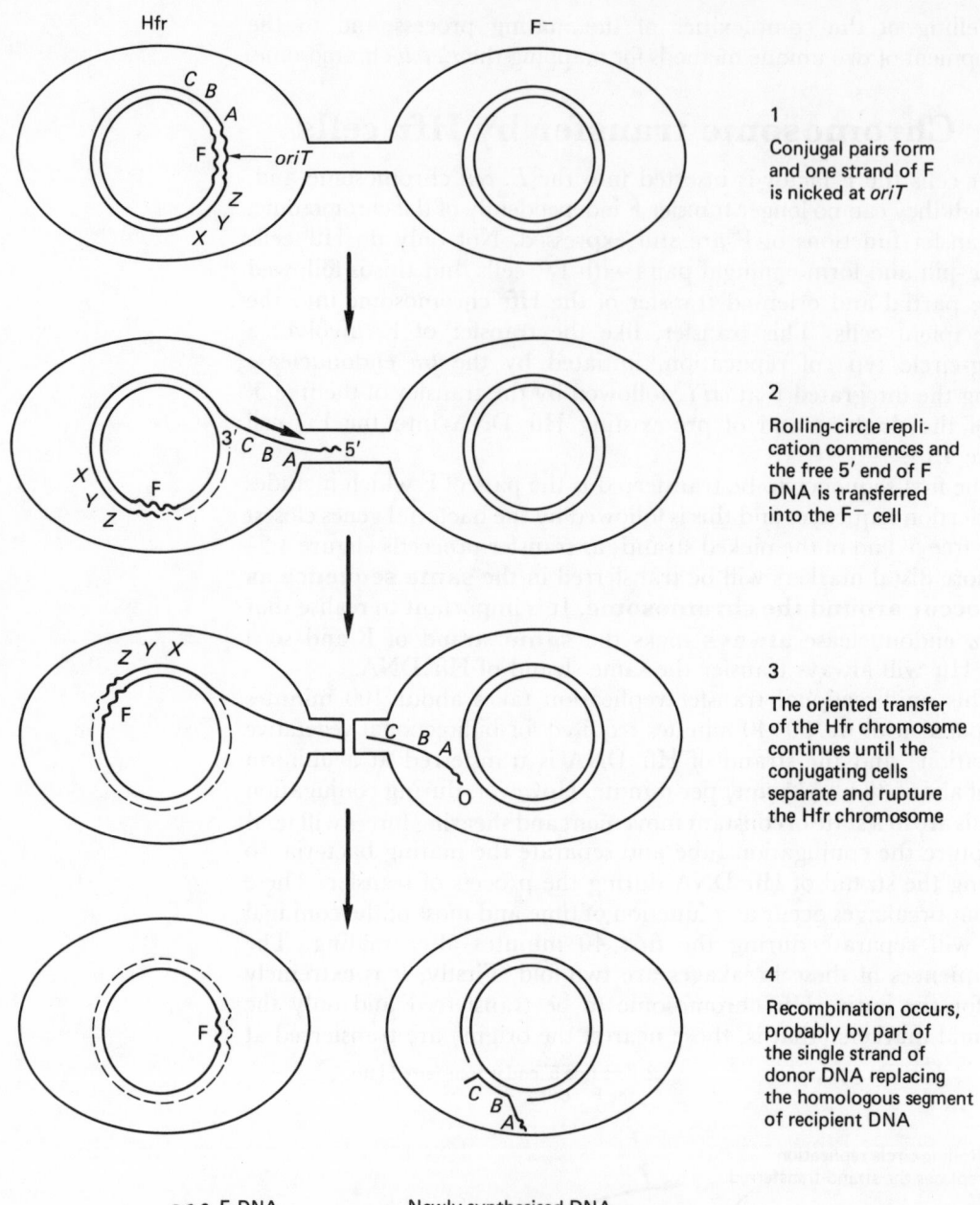

1
Conjugal pairs form
and one strand of F
is nicked at oriT

2
Rolling-circle repli-
cation commences and
the free 5' end of F
DNA is transferred
into the F⁻ cell

3
The oriented transfer
of the Hfr chromosome
continues until the
conjugating cells
separate and rupture
the Hfr chromosome

4
Recombination occurs,
probably by part of
the single strand of
donor DNA replacing
the homologous segment
of recipient DNA

∿∿ F DNA   --- Newly synthesised DNA

Figure 4.2
Hfr transfer

The first sequences to be transferred
are part of F followed by the
sequential transfer of the bacterial
genes A...B...C...; the remainder
of F is only transferred in the rare
instances where the entire
chromosome has been transferred.

high frequency. Secondly, the further a given Hfr marker is from the
origin of transfer the later is the time it is transferred and so the greater
is the probability that shearing will separate the Hfr and F⁻ cells before
that marker can enter the F⁻ cell; hence, the nearer an Hfr marker
is to the origin the more frequently it will be transferred to an F⁻ cell.

The zygotes formed are partial diploids and they contain an intact
F⁻ chromosome and variable segments of the Hfr chromosome, all
including the origin. Recombination can now occur and generate a

variety of stable recombinants. This recombination is $recA^+$ dependent and involves homologous pairing between the corresponding nucleotide sequences on the donor and recipient molecules, strand separation of the recipient DNA duplex and the formation of a molecule of heteroduplex DNA. If, as is most likely, the Hfr DNA remains single-stranded after transfer, then recombination probably occurs in the same way as has been suggested for the integration of transforming DNA (figure 2.13). Within the resulting heteroduplex segment, one strand of F^- DNA is base-paired with the homologous segment of Hfr DNA and, after one round of semi-conservative replication, this heteroduplex will segregate one recombinant and one parental type chromosome.

The recombinants from Hfr × F^- crosses are nearly always F^-; this is because the rupture of the conjugation tube during transfer usually prevents the extreme distal end of the Hfr chromosome from entering the F^- cell. As a consequence only part of F is transferred; it is only when an extreme distal marker (z^+) is selected that both parts of F are transferred and Hfr recombinants can be recovered.

This unique method of transfer allows the Hfr chromosome to be mapped by two novel methods — mapping by gradient of transmission and mapping by interrupted mating.

4.3 Mapping by gradient of transmission

Conjugation is usually carried out by mixing liquid cultures of genetically different Hfr and F^- strains and allowing the cells to form Hfr–F^- conjugal pairs. The mating mixture is then gently diluted and plated on a suitable solid medium to select the desired class of recombinant; this prevents the formation of any new Hfr–F^- contacts but does not separate any existing conjugal pairs. After the recombinants have grown into colonies, each one is tested to see which other donor markers have been inherited. Hfr transfer is a very efficient process and when a proximal marker is selected there may be one recombinant for every 10 Hfr cells plated; this is several orders of magnitude higher than the frequency of recombinants in P1 transductions.

The oriented and partial transfer of the Hfr chromosome produces a population of merozygotes containing variable pieces of the Hfr chromosome, all commencing at the origin but terminating at random points along the length of the chromosome. The result is that the nearer a marker to the origin, the more often it is transferred into a merozygote — in other words there is a gradient of marker transmission. Furthermore, once introduced into an F^- cell every marker has the same chance of being recombined into the F^- chromosome and so the frequency with which a particular marker appears among the recombinants is directly proportional to its frequency of transfer.

Mapping by gradient of transmission is best illustrated by considering a cross carried out by **Wollman, Jacob and Hayes** using HfrH, which transfers its chromosome in the sequence

$$\text{origin} \quad \longleftarrow \quad \overline{\textit{thr leu azi ton lac gal } (\lambda) \textit{ mal str xyl mtl}}$$

In their first experiment (table 4.1) they selected thr^+ leu^+ str^r recombinants by plating on minimal medium containing streptomycin (on this medium the F$^-$ parent could not grow as it required both threonine and leucine, while the Hfr parent was killed by the streptomycin), and each thr^+ leu^+ recombinant was tested to see which other donor markers it had inherited. They found that the unselected markers formed a gradient of recovery from azi^s (92 per cent) to $(\lambda)^-$ (15 per cent) so that any other marker lying within this segment could easily be assigned to a precise position on the genetic map simply by determining how frequently it appeared among the thr^+ leu^+ recombinants. Note that the proximal markers thr^+ and leu^+ are so closely linked that, in these experiments, they were selected as if they were a single gene.

When mapping by gradient of transmission, selection **must** be made for a proximal marker as then the results will be relatively unbiased by recombination and the observed gradient will depend on the occurrence of random chromosome breakage during transfer. However, if a distal marker is selected, then every merozygote will receive a segment of the Hfr chromosome extending from the origin to beyond the selected marker and, within this segment, the frequency of marker recovery will depend solely upon the frequency of recombination; since this is the same for all markers, no deductions about gene order can be made for markers located within this segment. This is illustrated by the second cross, when gal^+ str^r recombinants were selected; all the donor markers located between the origin and gal^+ appear among the recombinants with the same frequency.

Table 4.1
Mapping by gradient of transmission

thr^+ leu^+ str^r recombinants were selected by plating on minimal medium containing streptomycin, and gal^+ str^r recombinants selected on minimal medium containing galactose (instead of glucose), threonine, leucine and streptomycin. $(\lambda)^+$ and $(\lambda)^-$ indicate the presence or absence of a λ prophage at $att\lambda$.

Data from Wollman, E. L., Jacob, F. and Hayes, W., *Cold Spring Harbor Symp. Quant. Biol.*, **21**, 141 (1956).

HfrH thr^+ leu^+ azi^s ton^s lac^+ gal^+ $(\lambda)^-$ mal^+ str^s xyl^+ mtl^+
F$^-$ thr^- leu^- azi^r ton^r lac^- gal^- $(\lambda)^+$ mal^- str^r xyl^- mtl^-

Selected markers	Percentage frequency of unselected Hfr markers among the selected recombinants								
	thr^+ leu^+	azi^s	ton^s	lac^+	gal^+	$(\lambda)^-$	mal^+	xyl^+	mtl^+
thr^+ leu^+ str^r	—	92	73	49	31	15	0	0	0
gal^+ str^r	75	75	74	74	—	84	0	0	0

The mal^+ xyl^+ and mtl^+ markers do not appear among the recombinants for two reasons. Firstly, they are a long way from the origin and so are only very rarely transferred into the F^- recipient. Secondly, they are closely linked to the str^s gene and many of the recombinants inheriting the mal^+, xyl^+ and mtl^+ genes would also inherit str^s from the Hfr parent; these recombinants would be sensitive to streptomycin and killed by the streptomycin in the medium.

4.4   Mapping by interrupted mating

This method is similar to mapping by gradient of transmission but at various times after mixing the Hfr–F^- conjugal pairs are artificially separated by violent agitation in a high-speed blender; the mating mixture is immediately diluted to prevent the reforming of conjugal pairs, plated to select recombinants for a proximal marker and each recombinant tested to see which unselected donor markers it has inherited. This method establishes the time at which a marker first enters the F^- cells, making it possible to construct a genetic map with the distances expressed in time units. The map position of an unmapped marker is easily determined by establishing the time at which that marker first appears among the recombinants, and it is the method most frequently used for mapping by conjugational analysis.

The method is possible because the Hfr and F^- cells form conjugal pairs almost instantly after mixing and because the oriented transfer of the chromosome takes place at a constant rate — consequently, a given marker will always enter the F^- cell at the same time after mixing the Hfr and F^- cells.

Some of the results of a cross made by **Jacob and Wollman** are shown in table 4.2 and figure 4.3. This cross was the same as in the previous experiment except that the mating mixture was violently

---

| HfrH | thr^+ leu^+ azi^s ton^s lac^+ gal^+ str^s |
| F^- | thr^- leu^- azi^r ton^r lac^- gal^- str^r |

| | Constitution (*per cent*) of thr^+ leu^+ recombinants | | | | | | | | | |
| | *Untreated samples* | | | | | *Blended samples* | | | | |
Time sampled (*minutes*)	thr^+ leu^+	azi^s	ton^s	lac^+	gal^+	thr^+ leu^+	azi^s	ton^s	lac^+	gal^+
5	100	90	73	34	17	0	0	0	0	0
10	100	89	74	38	18	100	12	3	0	0
15	100	90	75	32	19	100	70	31	0	0
20	100	91	74	34	18	100	88	71	12	20
40	100	90	80	42	19	100	90	75	38	20

Table 4.2
Mapping by interrupted mating

At various times after mating, one sample was diluted and plated on minimal medium containing streptomycin; a second sample was vigorously blended for two minutes before plating. The recombinants were scored for the unselected markers.

Data from Jacob, F. and Wollman, E. L., *Sexuality and the Genetics of Bacteria*, Academic Press, New York (1961).

Figure 4.3
Marker recovery after interrupted mating

Each Hfr marker has a unique time of entry and a different level of incorporation.

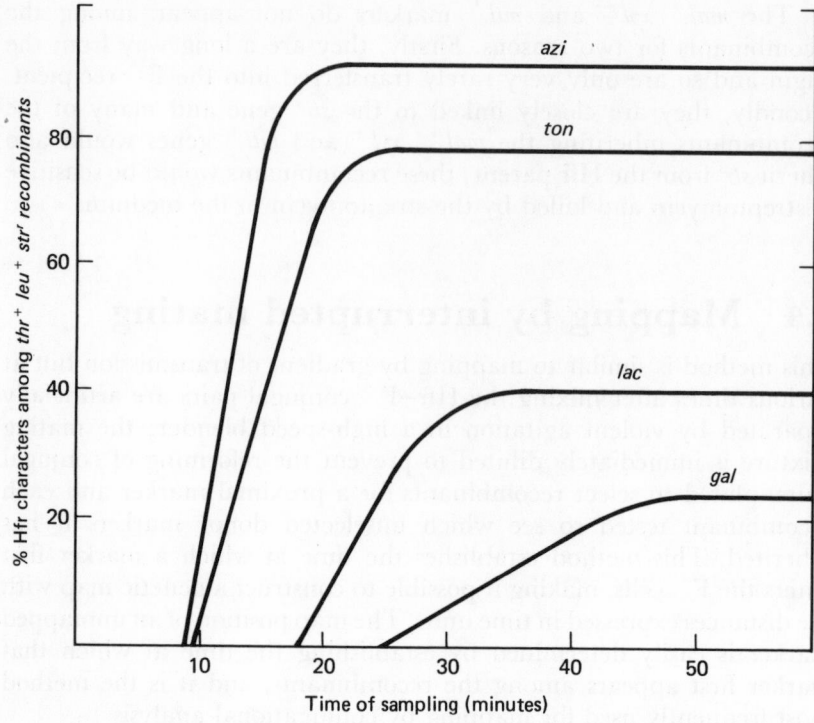

Figure 4.4
A comparison of genetic mapping by gradient of transmission and interrupted mating

The map distances expressed in kilobase pairs are calculated from the differences in the times of entry, on the basis that the chromosome is transferred at a uniform rate of 38 kb per minute.

agitated at various times after mating, diluted and then plated to select thr^+ leu^+ str^r recombinants.

The features to note are: (i) each marker first appears among the recombinants (and so first enters the F⁻ cell) at a particular time — when mating is interrupted before this time, the marker does not appear among the recombinants; (ii) each marker has a different level of incorporation (figure 4.3) and the earlier a marker is first transferred the higher is the plateau and the earlier it is reached — this is a consequence of the gradient of marker recovery; (iii) there is complete agreement between the maps constructed by interrupted mating and by gradient of transmission (figure 4.4).

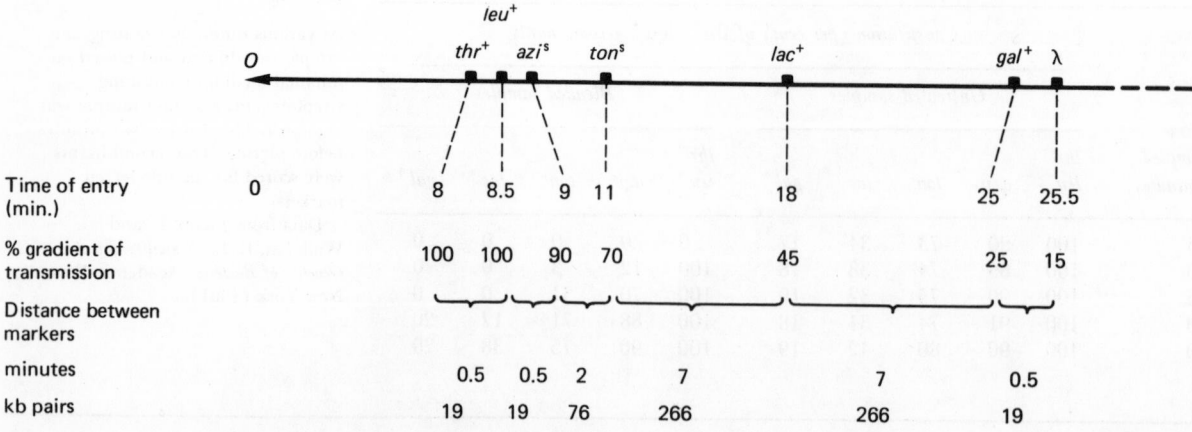

O		thr^+	azi^s		ton^s		lac^+		gal^+	λ
Time of entry (min.)	0	8	8.5		9	11	18		25	25.5
% gradient of transmission		100	100		90	70	45		25	15
Distance between markers										
minutes			0.5	0.5	2	7		7		0.5
kb pairs			19	19	76	266		266		19

4.5 The linkage map of *E. coli*

Most Hfr strains only transfer a segment of their chromosome at high frequency and so the use of a particular Hfr only enables about one-quarter of the *E. coli* chromosome to be mapped. However, there are a number of different Hfr strains, each produced by the integration of F at a different site on the chromosome, some transferring in a clockwise direction and others in an anti-clockwise direction; by using several Hfrs which transfer different but overlapping segments (figure 4.5), it is a simple matter to construct a composite map. By 1984 over 1100 genes had been identified and mapped, and a map showing some of these genes is shown in figure 4.5; this map is calibrated from 0 to 100 minutes (the time to transfer the entire chromosome) and the *thr* gene at 12 o'clock is arbitrarily assigned the position of 0 (and 100) minutes. All the genes described in this book are listed in table 4.3.

Mapping by gradient of transmission and by interrupted mating are most useful when mapping genes that are not very closely linked and more than 2 minutes apart on the genetic map. When mapping very closely linked genes or establishing the order of different mutant sites within the same gene (fine structure mapping), it is necessary to use a more conventional recombination analysis. It is important to

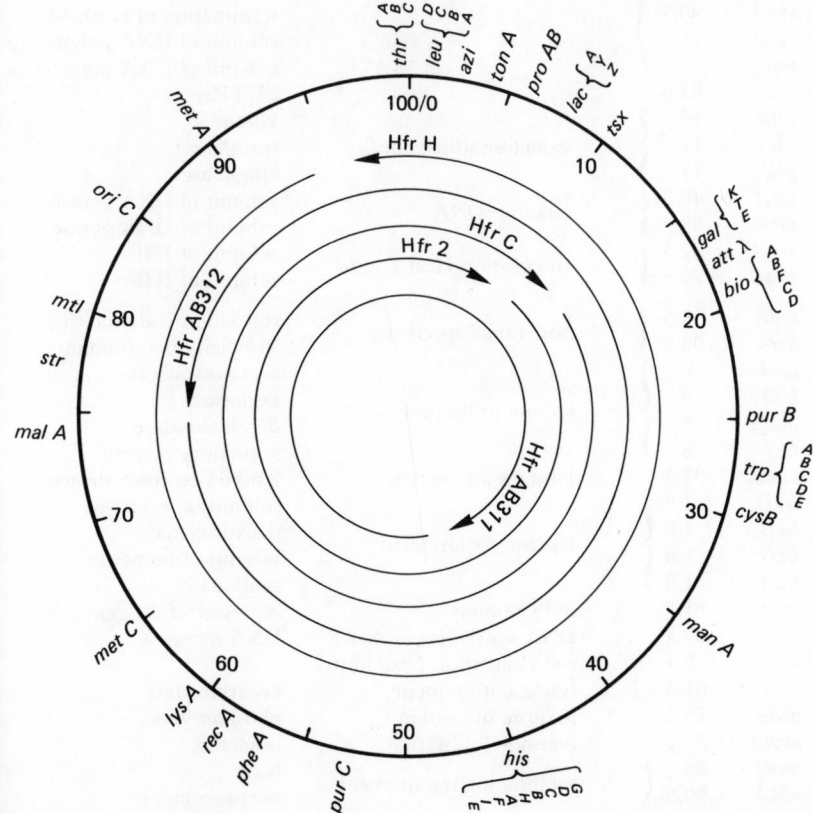

Figure 4.5
The linkage map of the *E. coli* chromosome

The outer circle shows the map positions of some of the many known genes, and a time scale based on an arbitrary origin at *thr*. The inner circles show the points of entry and the directions of transfer of five different Hfr chromosomes.

Table 4.3
Some important *E. coli* genes

Showing the map position, function affected and an indication of the type of activity of the protein product (if any) for each gene referred to in the text. Within each operon the genes are listed in map sequence, otherwise they are in alphabetical order. The arrows show the directions of transcription (↑ anticlockwise, ↓ clockwise) within the operons.

Adapted from Bachmann, B. R., *Microbiol. Rev.*, **47**, 180 (1983), which reproduces the current edition of the *E. coli* K-12 linkage map.

Gene symbol	Map position (minutes)	Character affected	Type of protein activity
araD	1.4	arabinose utilisation	epimerase
araA	1.4		isomerase
araB	1.4		ribulokinase
araC	1.4		regulatory protein
attλ	17.4	site for λ integration	none
attφ80	27.7	site for φ80 integration	none
azi	2.4	sodium azide resistance	
bioA	17.5	biotin requirement	
bioB	17.5		
bioF	17.5		
bioC	17.5		
bioD	17.5		
cysB	28	cysteine requirement	regulatory protein
dam	74.3	methylation	DNA adenine methylase
dcm	43	methylation	DNA cytosine methylase
din	—	damage inducible (SOS) genes	
dnaA	83	DNA synthesis	initiation
dnaN	83		subunit of DNA polymerase I
dnaB	91.7		chain elongation
dnaC	98.7		initiation & elongation
dnaG	67		DNA primase
dnaP	85.2		initiation
dnaT	98.7		termination of synthesis
dnaX	11		subunit of DNA polymerase I
dnaZ			subunit of DNA polymerase I
dut	81.6		dUTPase
galK	17	galactose utilisation	kinase
galT	17		transferase
galE	17		epimerase
gyrA	48.2	relaxing DNA	subunit of DNA gyrase
gyrB	83		subunit of DNA gyrase
himA	37.5	Integration Host Factor	subunit of IHF
himD	20.3		subunit of IHF
hsdR	98.5	host range specificity	restriction endonuclease
hsdS	98.5		specificity determinant
lacA	8	lactose utilisation	acetyltransferase
lacY	8		permease
lacZ	8		β-galactosidase
lacI	8		regulatory protein
lamB	91.5	lambda adsorption	lambda receptor protein
leuD	1.8	leucine requirement	subunit of isomerase
leuB	1.8		dehydrogenase
leuC	1.8		subunit of isomerase
leuA	1.8		synthetase
lexA	91.7	SOS response	repressor of *din* genes
lig	52.2	DNA synthesis (= *dnaL*)	DNA ligase
lon	9.7	cell elongation (*long* form)	
lysA	61.3	lysine requirement	decarboxylase
malA	75.2	maltose utilisation	phosphorylase
manA	35.7	mannose utilisation	isomerase
metC	65	methionine requirement	lyase
metA	90.5		*trans*-succinylase

mtlC	c.80.7	⎫	regulatory protein
mtlA	c.80.7	⎬ mannitol utilisation	transferase
mtlD	c.80.7	⎭	dehydrogenase
oriC	84	replication origin	none
pheA	56.6	phenylalanine requirement	dehydrogenase
phrA	16.2	⎫ photoreactivation	photolyase component
phrB	16.2	⎭	photolyase component
polA	86.5		DNA polymerase I
polB	2.0	⎬ DNA polymerisation	DNA polymerase II
polC	4.3		DNA polymerase III subunit
proA	5.8	⎫ proline requirement	reductase
proB	5.8	⎭	kinase
purB	25.2	⎫ purine metabolism	lyase
purC	53.3	⎭	synthetase
pyrF	28.3	pyrimidine metabolism	decarboxylase
recA	58.2		RecA protein
recD	60.7		subunit of exonuclease V
recB	60.7		subunit of exonuclease V
recC	60.7	general recombination	subunit of exonuclease V
recE	29.7		exonuclease VIII
recF	83		?
recG	c.82	⎭	?
rep	84.6	DNA melting	
rpoA	72.4	⎫	RNA polymerase α subunit
rpoB	90	⎬ transcription	RNA polymerase β subunit
rpoC	90		RNA polymerase β' subunit
rpoD	67	⎭	RNA polymerase σ subunit
sbcA	30.5	suppressor of *recBC⁻*	?
sbcB	43.8	suppressor of *recBC⁻*	exonuclease I
strM	76.8	resistance to streptomycin	
supC	27.2	ochre suppressor in *tyrT*	
thrA	0	⎫	dehydrogenase
thrB	0	⎬ threonine requirement	kinase
thrC	0	⎭	synthetase
tonA	4	⎫ phage T1 resistance	T1 receptor protein
tonB	27.7	⎭	
trpA	28	⎫	synthetase A protein
trpB	28		synthetase B protein
trpC	28	tryptophan requirement	isomerase + synthetase
trpD	28		transferase
trpE	28	⎭	synthetase
tsx	9.5	phage T6 resistance	T6 receptor protein
tyrT	27.2	structural gene for tyrosine tRNA1	none
umuC	26	UV-induction of mutations	?
ung	56	DNA repair	uracil−N−glycosylase
uvrA	92	⎫	subunit of *uvrABC* endonuclease
uvrB	17.6	⎬ excision repair	subunit of *uvrABC* endonuclease
uvrC	42		stimulates *uvrABC* endonuclease activity
uvrD	85.2	⎭	?
xseA	54	repair, degradation of single-stranded DNA	exonuclease VII
xylA	80	⎫	isomerase
xylB	80	⎬ xylose utilisation	kinase
xylR	80	⎭	regulatory protein

realise that conjugation, like transduction and transformation, is another method for producing partial diploids and that, when very closely linked genes are involved, two-point and three-point test cross data can be analysed in the same way as transduction data (section 2.3). This is possible because the gradient of transmission has little effect on groups of such closely linked markers and the same small segment of the Hfr chromosome is present in the majority of those merozygotes that produce the selected class of recombinant. However, when the markers are less than 2 minutes apart they can usually be co-transduced by phage P1 and it is usually more expedient to use transductional analysis.

4.6   Genetic analysis using F' strains

F-prime factors (section 3.2.5) arise spontaneously in populations of Hfr cells, and an F' carrying a particular bacterial gene can be isolated so long as a suitable Hfr is available. For example, in Hfr2 (figure 4.5) F is inserted between lac^+ and pro^+, the chromosome is transferred in the sequence

$$\text{origin} \leftarrow pro^+ - leu^+ \cdots gal^+ - lac^+$$

and an $F'lac^+$ is simply isolated. An Hfr2 lac^+ str^s donor and an $F^- lac^-$ str^r recipient are mixed and conjugation is interrupted after 30 minutes, so preventing chromosomal transfer of the lac^+ gene, and rare lac^+ str^r exconjugants are selected. Under these conditions, the only lac^+ exconjugants will be F^- cells that have received a rare spontaneously arising $F'lac^+$ from the population of donor cells. A typical $F'lac^+$ contains about 50 kb of bacterial DNA but some contain very much more; F140, for example, contains over 500 kb of *E. coli* DNA, over one-eighth of the chromosome.

One of the principal uses of F' factors is for constructing partial diploids, making it possible to study the relationships and interactions between genes in the same way that they can be studied in higher organisms. To illustrate these uses, consider three examples involving the lactose operon of *E. coli* (see box 4.1).

1. It enables dominance relationships to be established. If an $F'lac^+$ is introduced into an F^- $lacZ$ mutant, the $F'lacZ^+/lacZ^-$ partial diploid has a Lac$^+$ phenotype; the $lacZ^+$ gene is expressed and is said to be **dominant** while the $lacZ^-$ gene is not expressed and is **recessive.**

2. It is the best method for carrying out complementation tests in *E. coli*. Complementation is a simple genetic test used to determine whether two mutants, usually having similar phenotypes, are mutant within the same gene or within different genes. This standard test can be used whenever it is possible to put into the same cell the two genetic elements carrying the mutant genes. Consider when an $F'lac$ with a mutation in the $lacZ$ gene is introduced into an F^- $lacY$ mutant. The F' has a functional $lacY^+$ while the chromosome has a functional $lacZ^+$ gene and these can

Box 4.1 The lactose operon of *E. coli*

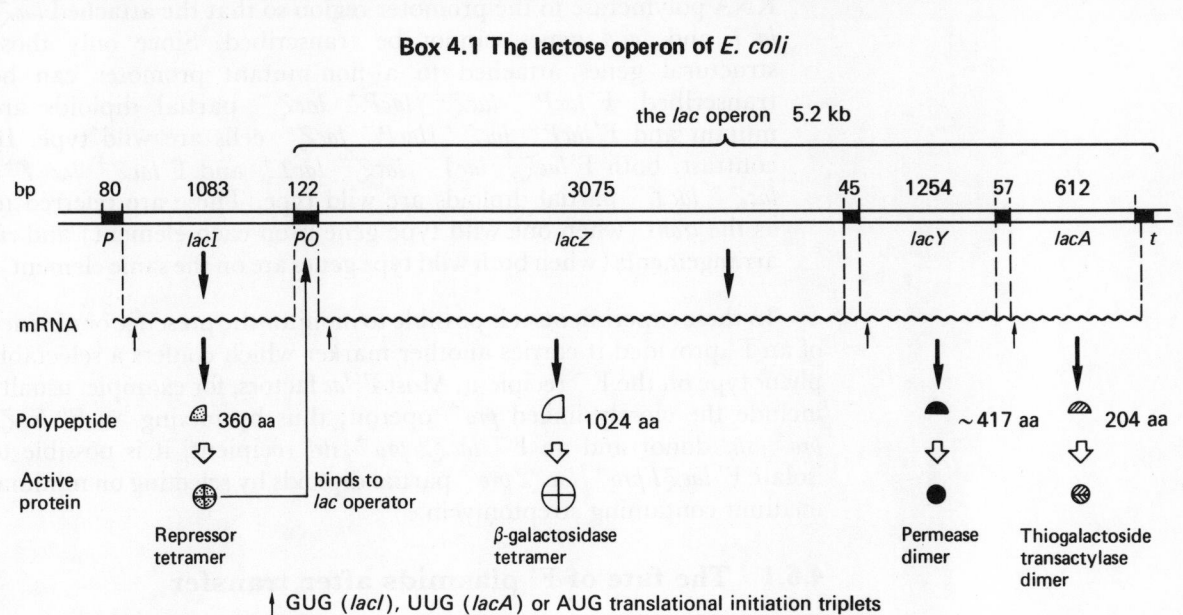

| GUG (*lacI*), UUG (*lacA*) or AUG translational initiation triplets

The lactose operon of *E. coli* consists of three structural genes under coordinate control. Transcription commences at a promoter (*lacP*) to the left of *lacZ* and transcribes a 5.2 kb messenger, ending at a terminator (*t*) beyond *lacA*. The three genes on the polygenic messenger are separately translated and the products form β-galactosidase which cleaves lactose into galactose and glucose, a permease required for transporting lactose across the cell membrane and a transacetylase whose function is not fully understood.

The operon is under the control of the adjacent *lacI* gene, encoding the lactose repressor. In the absence of allolactose, the inducer of the *lac* operon, the repressor tetramer binds to the *lac* operator (*lacO*) and prevents RNA polymerase from transcribing the operon. However, when allolactose is present it binds to the repressor; this prevents repressor from binding to *lacO* and permits RNA polymerase to bind to *lacP* and to initiate transcription.

The *lacI* gene is transcribed constitutively from its own promoter.

be transcribed and translated, producing β-galactoside permease and β-galactosidase respectively; since both gene products are present, the F'*lacZ⁻ lacY⁺*/*lacZ⁺ lacY⁻* partial diploid will have a wild phenotype. Thus when the mutations are in different genes the defective gene on one element is made good, or **complemented**, by the corresponding wild type gene on the other element and vice versa. On the other hand, when both mutations are in the same gene, as in a F'*lacZ1*/*lacZ2* partial diploid, neither element can produce β-galactosidase and the cell will remain lactose non-fermenting.

3. Mutation in genetic control sites (promoters, operators, etc.) can be distinguished from mutation in structural genes by use of the *cis–trans* test. Some Lac⁻ mutants are the result of mutations in the lactose promoter (*lacP⁻*); these interfere with the binding of

RNA polymerase to the promoter region so that the attached *lacZ*, *lacY* and *lacA* genes cannot be transcribed. Since only those structural genes attached to a non-mutant promoter can be transcribed, F'*lacP⁻ lacZ⁺/lacP⁺ lacZ⁻* partial diploids are mutant and F'*lacP⁺ lacZ⁺/lacP⁻ lacZ⁻* cells are wild type. In contrast, both F'*lacZ⁺ lacY⁻/lacZ⁻ lacY⁺* and F'*lacZ⁺ lac Y⁺/lacZ⁻ lacY⁻* partial diploids are wild type. These are referred to as the *trans* (when one wild type gene is on each element) and *cis* arrangements (when both wild type genes are on the same element).

In these experiments it is possible to monitor the presence or absence of an F' provided it carries another marker which confers a selectable phenotype on the F⁻ recipient. Most F'*lac* factors, for example, usually include the closely linked *pro⁺* operon; thus by mixing an F' *lacZ1 pro⁺/strˢ* donor and an F⁻ *lacZ2 pro⁻/strʳ* recipient, it is possible to isolate F' *lacZ1 pro⁺/lacZ2 pro⁻* partial diploids by selecting on minimal medium containing streptomycin.

4.6.1   The fate of F' plasmids after transfer

Once an F' has been transferred into an F⁻ cell it usually replicates autonomously and in synchrony with the bacterial chromosome, just like a wild type F. Occasionally, however, it may recombine with the bacterial chromosome in one of the following ways:

(i)   Reciprocal genetic exchange can occur between the bacterial genes on the F' and the homologous region of the bacterial chromosome. For example, reciprocal recombination could convert an F' *lac⁺/lac⁻* cell to F' *lac⁻/lac⁺*; this is equivalent to the formation of a stable transductant (figure 2.2).

(ii) The F' may integrate into the recipient chromosome to form an Hfr strain. This usually occurs because of homology between the bacterial genes on the F' and the chromosome and so the majority of derived Hfr strains will transfer the same chromosome segment in the same orientation. Thus an F' *lac⁺* would preferentially integrate into the *lac* region of the recipient chromosome.

(iii) Less frequently, F' can integrate in the same way as F because of homology between insertion sequences present both on F' and the chromosome. In the case of an F'*lac⁺* this would generate an Hfr chromosome containing two sets of the *lac* genes, one set at the standard map position and another set transposed to a new position at the site of F' integration.

Exercises

4.1   What is a merozygote? What genetic mechanisms can produce merozygotes of *E. coli*?

4.2 Why are the recombinants from Hfr × F$^-$ crosses nearly always F$^-$ whereas the exconjugants from F$^+$ × F$^-$ crosses are usually F$^+$?

4.3 Distinguish between mapping by gradient of transmission and mapping by interrupted mating.

4.4 In Hfr × F$^-$ crosses only bacterial genes close to and on one side of the integrated F plasmid can be transferred at high frequency. Why?

4.5 Explain how Hfr strains are formed. Why is it that different Hfrs have different origins and directions of transfer?

4.6 Compare and contrast the transfer of F-DNA in F$^+$ × F$^-$ and Hfr × F$^-$ matings.

4.7 You have isolated a new Lac$^-$ mutant (*lac-1*) of *E. coli*. Show how you would determine whether the *lac-1* mutation occurred within a structural gene or within a genetic control site.

4.8 You have isolated a further Lac$^-$ mutant (*lac-2*) and have shown that both the *lac-1* and *lac-2* mutations occurred within structural genes. How can you show whether the mutant sites are within the same gene or in different genes?

4.9 What two simple methods would enable you to distinguish whether a recombinant from an Hfr × F$^-$ cross was F$^-$ or F′.

4.10 A new mutation has been isolated in *E. coli* and mapped by an interrupted mating experiment. Is this technique adequate for accurate genetic mapping? Could the resolution of the mapping have been increased by a P1-mediated transduction experiment?

4.11 In the cross HfrH *lac-1 purE*$^+$ *str^s* × F$^-$ *lac-2 purE*$^-$ *str^r* purine-independent recombinants were selected on minimal medium containing streptomycin; 7 per cent of these recombinants were able to ferment lactose. In the reciprocal cross HfrH *lac-2 purE*$^+$ *str^s* × F$^-$ *lac-1 purE*$^-$ *str^r* 75 per cent of the Pur$^+$ recombinants were also Lac$^+$. How do you explain these results and what is the order of the *lac-1* and *lac-2* mutations relative to *purE*?

4.12 You have isolated a new Hfr which transfers in the sequence

$$O-lac-gal-trp \ldots pro$$

and in Hfr × F$^-$ crosses you recover a few *pro*$^+$ exconjugants, even when mating is interrupted after only 10 minutes. How can you explain this unexpected result and how would you confirm your deductions?

4.13 You have isolated a new Hfr strain from an F$^+$ *trp*$^-$ strain of *E. coli*. This Hfr was mated to an F$^-$ *lac*$^-$ *mal*$^-$ *pro*$^-$ *pheA*$^-$ *leu*$^-$ *metA*$^-$ *his*$^-$ *lysA*$^-$ strain for 100 minutes and (a) *lac*$^+$ and (b) *metA*$^+$ recombinants selected on appropriately supplemented media. Counterselection against the Hfr was made by omitting tryptophan from the media. The percentage of recombinants inheriting each of the unselected donor markers was determined by testing samples of the recombinants, with the following results:

Selected marker	% Recombinants inheriting the following unselected markers							
	lac$^+$	*mal*$^+$	*pro*$^+$	*pheA*$^+$	*leu*$^+$	*metA*$^+$	*his*$^+$	*lysA*$^+$
lac$^+$	—	15	95	0	80	60	0	0
metA$^+$	45	40	52	0	48	—	0	15

Explain the differences between the crosses and deduce the origin and orientation of the Hfr donor.

4.14 In a cross between *E. coli* Hfr *gal*$^+$ *str*^{s} and F$^-$ *gal*$^-$ *str*^{r} strains carrying, respectively *m5 c*$^+$ *mi*$^+$ and *m5*$^+$ *c mi* λ prophages, *gal*$^+$ *str*^{r} recombinants were selected and analysed for the type of λ phage released on induction. 1540 recombinants carried parental type prophages and 60 carried recombinant phages as follows:

$$m5^+ \ c^+ \ mi^+ \quad \text{and} \quad m5 \ \ c \ mi \quad\quad 36$$
$$m5^+ \ c^+ \ mi^+ \quad \text{and} \quad m5 \ \ c \ mi^+ \quad\quad 19$$
$$m5 \ \ c^+ \ mi^+ \quad \text{and} \quad m5^+ c \ mi \quad\quad 5$$

What is the order of the markers in the prophage?
(Note: λ_{m5} produces medium-size plaques, λ_{mi} minute plaques and λ_c clear plaques)

References and related reading

Jacob, F. and Wollman, E. L., *Sexuality and the Genetics of Bacteria*, Academic Press, New York (1961).

Low, K. B., '*Escherichia coli* K-12 F-prime factors, old and new', *Biol. Rev.*, **36**, 587 (1972).

Wollman, E.L., Jacob, F. and Hayes, W., 'Conjugation and genetic recombination in *Escherichia coli* K-12', *Cold Spring Harbor Symp. Quant. Biol.*, **21**, 141 (1956).

RECOMBINATION 5

5.1 Introduction

General recombination, as distinct from site-specific recombination, occurs between two molecules of DNA with homologous nucleotide sequences (or, in eucaryotes between two homologous chromosomes) and results in the reciprocal exchange of DNA sequences between these molecules. The recombinational process is of fundamental importance as it is both the main tool of genetic analysis, permitting the construction of genetic (linkage) maps, and it is used, both in nature and by the experimentalist, to combine into a single genotype the most favourable combinations of genes; in the absence of recombination every chromosome would have a 'fixed' genetic content changeable only by mutation, and all the genes on it would show complete linkage.

The mechanisms of recombination are only partly understood and only recently has it become possible to define some of the events involved in molecular terms; although much of the genetic data which led to the molecular models for recombination derives from studies using ascomycetes, more recent molecular studies using *E. coli* and its phages make it certain that similar, even if not identical, mechanisms operate in procaryotes.

The first theory of recombination, the **chiasmatype theory**, was suggested by **F. A. Janssens** in 1909 even before the discovery of linkage. Later, during the 1930s, **Cyril Darlington** developed an intricate model to explain recombination occurring during meiosis in higher organisms. Both these models required two homologous but non-sister chromatids to break at exactly corresponding points along their length and to rejoin in a new combination (figure 5.1(A)); it was a necessary consequence of these models that recombination was always reciprocal and that, for every heterozygous marker, two of the meiotic products carried one allele and two the other. However, recombination is not always reciprocal and it does not always give this regular 2:2 segregation.

A rather different type of model was first suggested by **John Belling**

Figure 5.1
**Models for genetic
recombination**

(A) Breakage and reunion models
Recombination and replication are
separate events. Chiasmata are the
result of two non-sister chromatids
breaking and rejoining in a new
combination, at exactly homologous
positions.
(B) Copy choice models
Recombination is a consequence of
replication. Each member of a pair of
homologous chromosomes acts as a
template for a new daughter
chromosome; during replication the
daughter chromosomes can switch
from one parental template to the
other.

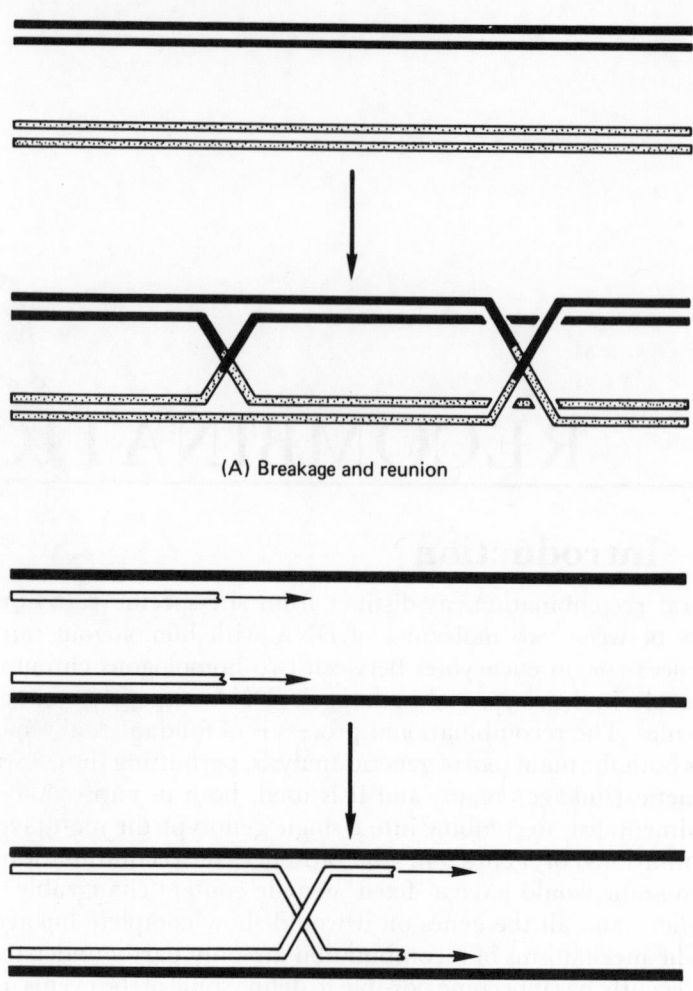

(A) Breakage and reunion

(B) Copy choice

in 1931 and revived by **Alfred Sturtevant** (1949) and **Joshua
Lederberg** (1955) in an attempt to explain recombination in
bacteriophage and bacteria. These **copy-choice** models suppose that
each parental chromosome acts as a template for the formation
of a new daughter chromosome and that during replication the
daughter chromosomes can switch from one parental template to the
other, so producing daughter chromosomes that are templated in part
against the maternal and in part against the paternal chromosome
(figure 5.1(B)). These models are unacceptable for two reasons. Firstly,
when multiple cross-overs occur they must always involve the newly
synthesised pair of non-sister chromatids, and the original parental
chromosomes will always be non-recombinant; and yet we know from
tetrad analysis that three-strand and four-strand double cross-overs
occur regularly. Secondly, when the model is translated into molecular
terms it requires the conservative replication of DNA, whereas we know
that DNA replicates semi-conservatively.

5.2   Aberrant segregations

The most precise information on the consequences of individual recombination events comes from tetrad analysis in ascomycetes, where the four products of an individual meiosis can be isolated and analysed. These analyses have revealed three features of the recombination process. Firstly, recombination is nearly always reciprocal and each tetrad shows regular 2:2 segregation of every heterozygous marker. Secondly, each recombination event (or cross-over) involves only two of the four meiotic products, and so only two of the four chromatids, and these products carry reciprocally exchanged segments. Thirdly, multiple recombination events may involve two, three or all four chromatids.

The first regular occurrence of abnormal segregation was reported in 1953 by **Carl Lindegren** in Bakers' yeast, *Saccharomyces cerevisiae*. In crosses between mutant and wild type strains he found some asci that contained (3 wild type + 1 mutant) or (1 wild type + 3 mutant) spores instead of the expected (2 wild type + 2 mutant). This non-reciprocal recombination event, occurring during meiosis and giving rise to a tetrad showing an abnormal segregation ratio, he called **gene conversion**, but the importance of this unexpected discovery was not realised until 1955, when **Mary Mitchell** showed not only that similar 3:1 (or rather 6:2) segregations occurred in crosses between pyridoxin-requiring mutants of *Neurospora crassa* but that a very closely linked site within the same locus continued to shown normal 2:2 segregation. This clearly demonstrated that the aberrant segregations were due to the abnormal behaviour of a particular mutant site and not to the abnormal behaviour of a whole chromosome.

Extensive studies with *Sordaria fimicola* have shown that about 0.2 per cent of all asci show either 6:2, 5:3 or irregular 4:4 segregation. Furthermore, about 30 per cent of these asci show crossing-over between closely linked flanking markers, and in nearly every instance the **same** two chromatids were involved in both gene conversion and reciprocal crossing-over.

Particularly interesting are the asci showing 5:3 segregation; segregation must have occurred at the mitotic division immediately following meiosis, implying that gene conversion had affected only one-half of a meiotic product or, in molecular terms, one strand of a duplex molecule of DNA.

Probably the most informative data on gene conversion have been collected by **Seymour Fogel and his colleagues** using *Saccharomyces cerevisiae*. In one series of experiments reported in 1969 they made pairwise crosses between a series of allelic mutants at the *ARG4* locus. (Note that yeast geneticists designate dominant wild type alleles by capital letters and recessive mutant alleles by lower case letters. *ARG4* refers to both the locus and the wild type allele; *arg1 arg2, arg4* and *arg17* are four different mutant sites within the locus.) These crosses were of the type *A ARG3 arg17 b* × *a arg3 ARG17 B* and all the asci were completely analysed. The data (table 5.1) show several features of importance.

Table 5.1
Abnormal segregation in
Saccharomyces cerevisiae

Fogel and Mortimer crossed four allelic arginine-requiring mutants in pairwise combinations and fully analysed all the asci by back-crossing with each of the parental strains. The table gives the numbers of abnormal asci showing conversion at the proximal (left-hand) site only, the distal site only and at both sites; conversion can occur either towards the mutant or towards the wild type allele. Note that in crosses between the most closely linked sites, more asci show gene conversion than show normal reciprocal recombination.

From Fogel, S. and Mortimer, R. K., *Proc. Natl. Acad. Sci. USA*, **62**, 96 (1969).

	Linkage order	—arg4–arg1–arg2–arg17—	
Cross	arg4 × arg17	arg1 × arg2	arg2 × arg17
Nucleotide pairs separating the mutant sites	1060	520	128
Asci analysed	697	502	544
Asci showing conversion at the proximal site only			
WT to mutant 1:3	5	3	3
mutant to WT 3:1	3	3	1
Asci showing conversion at the distal site only			
WT to mutant 1:3	20	11	2
mutant to WT 3:1	18	10	3
Asci showing conversion at both sites			
3:1 + 1:3	2	13	14
1:3 + 3:1	1	10	13
Asci showing abnormal segregations	49	50	36
Asci showing reciprocal recombination	9	5	0

First, over 7 per cent of the asci showed abnormal segregation. Second, the distal site within the *ARG4* gene (that is, the site farther from the centromere) was always converted more frequently than the proximal site; thus there was a **polarity** of conversion frequency from one end of the gene to the other. Third, some asci showed conversion at **both** sites of heterozygosity and they noted: (i) that the frequency of double site conversions increased as the distance between the mutant sites decreased; (ii) double site conversions were rare when the two sites of heterozygosity were more than 1000 base pairs apart; (iii) in each doubly converted ascus the marker combinations showed that the **same** chromatid was involved in **both** conversions; and (iv) about 50 per cent of the asci showing gene conversion were also recombinant for the flanking markers.

These results clearly demonstrated that conversion is frequently associated with crossing-over and involves a segment of the chromosome up to about 1000 base pairs long — thus when two mutant sites are very close together, they will frequently fall within the same segment of DNA and double site conversions will be frequent; when they are farther apart, only single site conversions will be recovered.

5.3 The molecular basis of genetic recombination

The results of tetrad analysis clearly show that although crossing-over is normally a reciprocal process, it occasionally, but nevertheless regularly, results in non-reciprocal recombination or gene conversion, a process that occurs **within** genes and involves finite **segments** of DNA. Gene conversion is a polarised process and frequently the same chromatid is involved both in gene conversion and normal reciprocal exchange. Furthermore, the asci showing 5:3 segregation make it clear that recombination within a gene can involve half-chromatids and produce a recombinant chromatid containing a short segment of hybrid genetic material; this segregates at the following mitosis by a process known as **post-meiotic segregation**.

Since all chromosomes, in both eucaryotes and procaryotes (with the exception of the RNA viruses), are single, very long molecules of DNA, it is clear that recombination must involve just one of the two strands making up a DNA molecule. In other words recombination generates a short segment of **hybrid** or **heteroduplex** DNA where one strand of the duplex is derived from one parental molecule and the second from the other parental molecule. In contrast, the pre-molecular theories of recombination assumed that recombination was a point phenomenon and cannot explain the aberrant segregations that result from gene conversion.

During the early 1960s the conclusions from studies on gene conversion independently prompted **Harold Whitehouse** and **Robin Holliday** to propose the first molecular models to account for recombination in eucaryotes. These models clearly defined recombination as an enzymatic process and were based on the assumption that recombination involved the breakage and re-union of **single** strands of DNA, and resulted in the formation of short segments of heteroduplex DNA. Since then, numerous alternative models have been suggested and the original Holliday model has been revised and refined as our knowledge of the recombination process has expanded. We will consider the present version of the Holliday model and a conceptually similar model proposed in 1975 by **Matthew Meselson and Charles Radding** to take into account some observations not easily reconcilable with the Holliday model.

5.4 The Holliday model for genetic recombination

According to this model, recombination between two closely paired homologous molecules of DNA is initiated by a nuclease recognising the homologous nucleotide sequences of these molecules and making single-strand nicks in non-sister strands of the **same** polarity; these nicks are made at identical positions on each parental molecule, and one end of each nicked strand now starts to **dissociate** from its intact sister strand (figure 5.2(1)). Each dissociated strand now **invades** the single-strand gap left in the homologous molecule and, because the two parental molecules are homologous and, because the two invading strands have the same polarity, the invading strands can be **assimilated** into the homologous molecules by the formation of H-bonds between the complementary (or nearly complementary) base sequences (figure 5.2(2)), the discontinuities remaining at the new single-strand joints eventually being sealed by ligation. These events generate a recombinational intermediate, made up of two molecules of DNA containing short reciprocal segments of heteroduplex DNA and connected by a **half-chiasma** (so-called because only two of the four strands of DNA are involved). However, this half-chiasma is not static and the rotational properties of the DNA molecule can drive it along the recombination intermediate. This **branch migration** involves further changes of pairing partners within the duplex DNA molecules and so extends the segments of heteroduplex DNA — in effect the half-chiasma is 'slid' along the recombinational intermediate (figure 5.2(3a)). In certain instances, it has been possible to isolate these

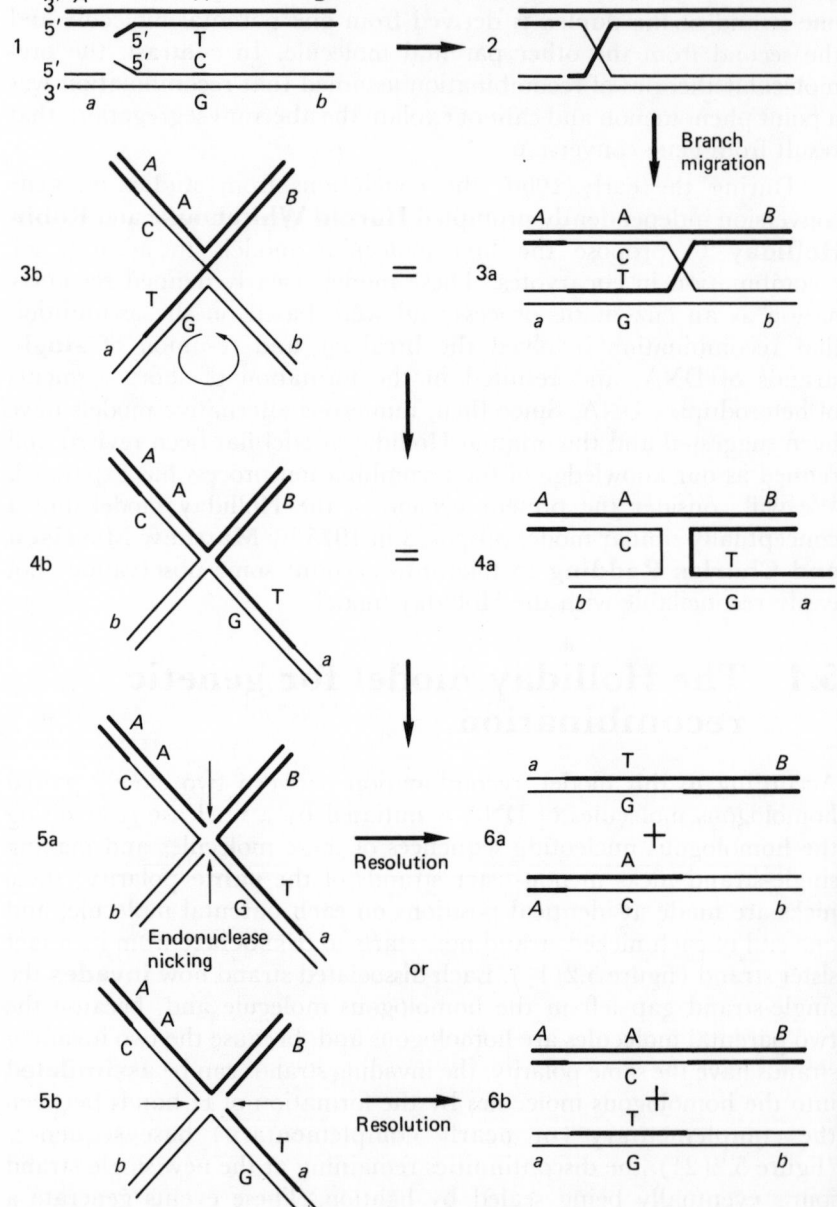

Figure 5.2
The Holliday model for recombination

The outcome of a cross $AMB \times amb$; M and m are represented by A–T and C–G base pairs respectively. Note that only the two chromatids participating in recombination are shown.

(1) **Breakage**   Strands of the same polarity are nicked at homologous positions.
(2) **Assimilation**   The nicked strands dissociate and the 5′ ends invade the single-strand gaps in the homologous parental molecules.
(3) **Branch migration**   The segments of heteroduplex DNA are extended by branch migration. In (3b) the joint molecule is redrawn in a planar configuration.
(4) **Isomerisation**   (4a) and (4b) are isomeric forms of the joint molecule shown in (3a) and (3b). Compare (3b) and (4b) with the observed chi-form intermediates shown in figure 5.6.
(5) **Resolution**   The joint molecule is nicked in either a N–S or an E–W plane to resolve it into two separate recombinant molecules.
(6) **Ligation**   The remaining nicks in the recombinant molecules are sealed by ligation.

Either type of resolution generates reciprocal heteroduplex molecules but only nicking in the N–S plane results in the reciprocal recombination of the flanking markers.

intermediates and when they are dropped on to a flat surface and examined by electron microscopy they are found to be X-shaped, and in figure 5.2(3b) the same intermediate is redrawn in this planar configuration; these structures are referred to as either **Holliday intermediates** or **chi-forms**.

Because DNA molecules are very flexible, these chi-forms can exist in several equivalent molecular forms or isomers. Thus, retaining A and B in position and rotating a over b in a vertical plane produces the symmetrical structure shown in figure 5.2(4), while the further rotation of B over a produces a chi-form in which the intact strands form the half-chiasma. The symmetrical chi-form is now easily resolved into two separate molecules by endonuclease nicking in either a N–S or E–W plane (figure 5.2(5a) and 5.2(5b)); finally, after the recombinant molecules have separated, the remaining single-strand gaps are sealed by ligation (figure 5.2(6)).

Resolution in either plane generates molecules containing reciprocal segments of heteroduplex DNA; when cleavage is in a N–S plane the markers flanking these segments are reciprocally recombinant (Ab and aB), but with E–W cleavage they are non-recombinant (AB and ab). At a molecular level the two types of resolution are identical and, as is observed, are expected to occur at comparable frequencies.

In figure 5.2(1) the parental chromosomes are shown as being heterozygous for a gene located between A and B. The two alleles are assumed to differ by a single base pair substitution mutation and the wild type allele (M) is represented by an A–T base pair and the mutant allele (m) by a C–G base pair; observe that recombination generates reciprocal heteroduplex segments, one with a T–G and the other with an A–C mismatched base pair. It is only these sites with mismatched base pairs that can show gene conversion, and the precise pattern of segregation will depend on the subsequent behaviour of these two pairs of mismatched bases.

5.4.1 Post-meiotic segregation and error correction

One of two things must now happen to each mismatched base pair (figure 5.3) Firstly the mismatch may persist until the next replication and then undergo post-meiotic segregation; the rules of complementary base pairing now apply and a heteroduplex with an A–C mismatch (for example) will be converted into two homoduplex molecules, one with an A–T (M) and the other with a G–C (m) base pair. Secondly, **error correction** or **mismatch repair** may occur. This is an enzymatic process and nucleases recognise the distortion in the double helix caused by the mismatch and excise a segment from **one** strand of the heteroduplex; the segment removed is about 1000 nucleotides long and spans the mismatch. The resulting single-strand gap is filled in by DNA polymerase inserting the correct complementary base and, finally, the discontinuities are sealed by ligation. Since either strand of the heteroduplex can be excised and replaced, an A–C mismatch can

(A) Post-meiotic segregation

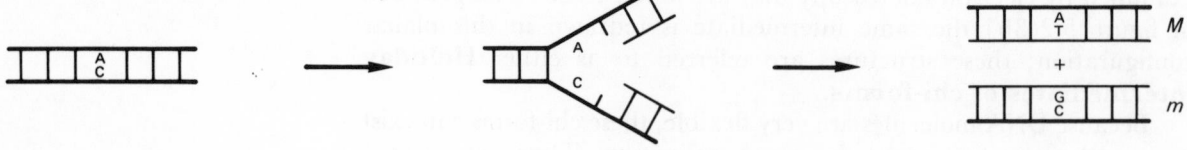

(B) Error correction

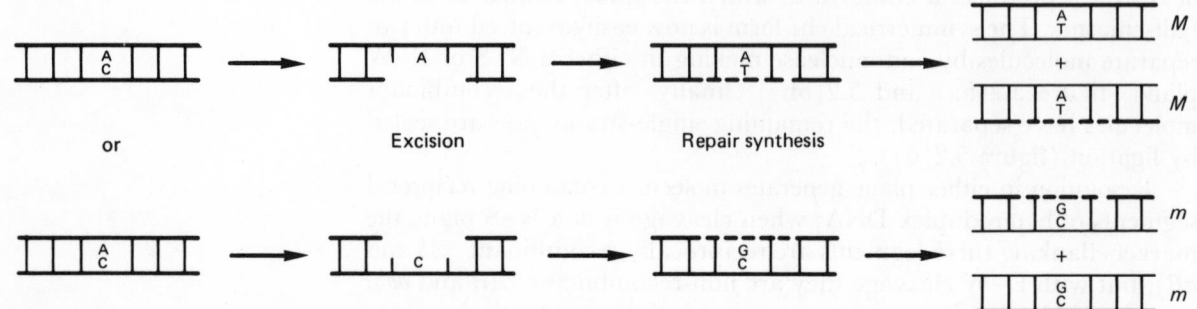

or   Excision   Repair synthesis

Figure 5.3
Post-meiotic segregation and error correction

The figure illustrates the possible consequences for a mismatched A–C base pair in a heteroduplex segment. At this position an A–T base pair represents the wild type (*M*) and a G–C base pair the mutant (*m*).

be corrected either to an A–T (*M*) or to a G–C (*m*) homoduplex molecule.

According to the Holliday model, recombination generates reciprocal heteroduplex segments. Since neither, either or both mismatches can be corrected, and since correction can be towards either the mutant or the wild type, a variety of abnormal asci showing gene conversion can be produced (table 5.2).

This type of model readily accounts for the polarity of gene conversion. Recombination is initiated by nucleases recognising specific nucleotide sequences and these will determine the position of the proximal end of each heteroduplex segment; however, the position of the distal end is determined by how far branch migration has proceeded at the time of resolution. Since branch migration is a random process and can occur in either direction, the heteroduplex segments will be of variable length, and the further a marker from the initiation sequence

Table 5.2
Error correction and post-meiotic segregation

Error correction of a mismatched base pair can be towards either wild type or mutant, and can affect either one or both mismatched base pairs. In the absence of error correction, post-meiotic segregation will occur. Different combinations of error correction and post-meiotic segregation produce asci showing different types of segregation.

	First mismatched base pair			*Second mismatched base pair*
	Corrected to wild type	Corrected to mutant	Not corrected	
	6:2	normal 4:4	5:3	Corrected to wild type
		2:6	3:5	Corrected to mutant
			aberrant 4:4	Not corrected

the less frequently will it be included in a segment of heteroduplex DNA. Since only markers falling within such a segment can show gene conversion, there will be a gradient of conversion frequencies with sites nearest the initiation sequence showing the highest frequencies of conversion. Likewise, if two or more closely linked sites fall within the same segment of heteroduplex DNA, a single error correction event can result in simultaneous gene conversion at each site of heterozygosity. This is precisely what was observed by **Fogel and Mortimer**.

5.4.2   Reciprocal and non-reciprocal heteroduplex segments

The Holliday model may be regarded as the prototype of molecular models for genetic recombination and is attractive because of its simplicity; however, there are some data which cannot be explained by the formation of reciprocal segments of heteroduplex DNA and which support the notion that heteroduplex segments are often non-reciprocal and occur in only one of the meiotic products. For example, certain abnormal asci can only be explained if reciprocal heteroduplex segments are formed, and in some ascomycetes, such as *Sordaria*, these asci are common and it is clear that heteroduplex DNA normally forms symmetrically. On the other hand, in *S. cerevisiae* the situation is quite different as these asci are extremely rare, suggesting that in yeast heteroduplex segments are formed asymmetrically. It was these and other similar observations that led **Meselson and Radding** to propose their model for recombination.

5.5   The Meselson and Radding model for genetic recombination

This model is largely based on the concepts of the Holliday model but only one strand is nicked and this, after dissociation, invades the homologous molecule. Once again, recombination occurs in two stages; firstly an asymmetric enzyme-driven phase which generates a hetero-duplex in only one of the two participating molecules and, secondly, a symmetric phase involving branch migration and producing reciprocal heteroduplex segments in both molecules.

The first event is the nicking of one molecule of parental DNA (figure 5.4(1)) followed by strand dissociation from the nick. As dissociation proceeds the gap produced is filled in by DNA polymerase activity and the dissociated single strand synapses with the homologous region along the other parental duplex (figure 5.4(2)). Meselson and Radding suggest that this strand extension physically **displaces** the nicked strand from the duplex. At this stage the two strands of the recipient molecule partly dissociate and the invading strand is **assimilated** into the molecule by complementary base pairing, forming a displacement loop or **D-loop** (figure 5.4(3)). The newly displaced strand of the D-loop is now cleaved by endonuclease and digested away by $5' \rightarrow 3'$ exonuclease activity; note that the length of

Figure 5.4
The Meselson and Radding model for recombination

(1) **Nicking** A single strand of one parental duplex is nicked.

(2) **Dissociation** The nicked strand dissociates and synapses with the homologous region along the other parental duplex; the gap is filled in by 5' to 3' DNA polymerase activity.

(3) **Assimilation** The invading strand is assimilated into the parental duplex forming a D-loop.

(4) **Digestion** The displaced strand of the D-loop is digested by nuclease activity. Note that a heteroduplex is formed in only *one* of the parental molecules.

(5) *Either* (A) isomerisation occurs (*A* is rotated over *a*) and converts the joint molecule into a form with a half chiasma *or* (B) the free 5' and 3' ends are ligated together to form a half chiasma.

(6) **Branch migration** Occurring from left to right, this generates reciprocal heteroduplex segments.

The joint molecules shown in (5) and (6) can at any time be resolved in the same way as Holliday intermediates.

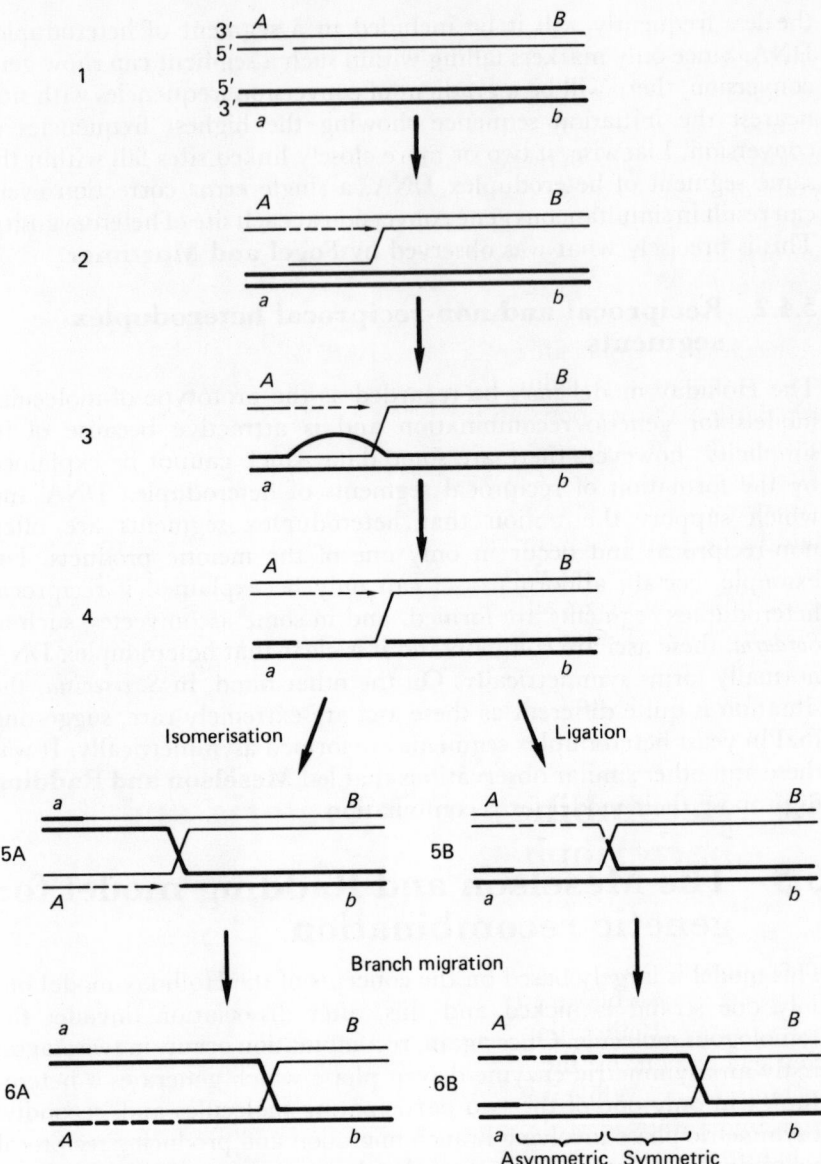

the heteroduplex segment that has been formed can be further extended by a continuation of this single-strand transfer, promoted by DNA polymerase activity along one parental strand and exonuclease activity along the other. This completes the asymmetric, enzyme-driven phase and the enzyme complex can now dissociate (figure 5.4(4)).

This joint molecule has only a single crossed-over strand but it can be converted into an alternative form with a half-chiasma either by isomerisation (retaining *B* and *b* in position and rotating *A* over *a* (figure 5.4(5A)), or by direct ligation of the two free ends (figure 5.4(5B)), and branch migration can then generate the isomerically equivalent intermediate shown in figure 5.4(6).

In molecular terms each of these intermediates is identical to the Holliday intermediate shown in figure 5.2(3) and can, at any time, be resolved to produce daughter molecules with either parental or recombinant associations of the flanking markers. Whereas the Holliday model results in two recombinant molecules containing reciprocally exchanged heteroduplex segments extending from the site of nicking to the site of resolution, the Meselson and Radding model generates an asymmetric heteroduplex segment (that is, in one molecule only) extending from the site of nicking to the site at which branch migration commences, and symmetrical heteroduplex segments (that is, in both molecules) from that site to the site of resolution. When a heteroduplex is formed in one daughter molecules only, certain types of aberrant segregation are not possible and so sites of heterozygosity near the site of nicking should not show these types of segregation. The rarity of these particular aberrant segregations in yeast can be explained if, in yeast, resolution occurs soon after completion of the asymmetric phase and before branch migration has proceeded to any significant extent. In *Sordaria*, however, these aberrant segregations are frequent and resolution must occur after branch migration has taken place.

Although this model does not require the **net** synthesis of DNA, it does involve the limited breakdown and resynthesis of DNA, and in several organisms there is evidence for a small amount of DNA synthesis during meiotic prophase.

5.6   Recombination in bacteria and bacteriophage

In bacteria and bacteriophage, tetrad analysis is not possible but, even so, the existence of both hybrid DNA and an error correcting mechanism have been deduced by using other methods; furthermore, it has been possible to isolate recombinational intermediates and to carry out recombination in *in vitro* systems so allowing a partial characterisation of the reactions involved in homologous genetic recombination. While it is unlikely that recombination follows exactly the same pathway in all organisms, the evidence, in general, favours a hybrid-DNA mechanism of the type suggested by Meselson and Radding. It is not possible to discuss this evidence in any detail but the following points are of particular interest.

(1)   In Hfr × F⁻ crosses of *E. coli*, only a single strand of donor DNA is transferred into the F⁻ recipient (section 4.2) and segments of this single strand are incorporated into the recipient DNA to produce heteroduplex segments. The simplest explanation is that the single-stranded donor DNA invades the duplex DNA of the recipient at a region of homology and displaces the recipient strand having the same polarity; this would be the equivalent of D-loop formation by the Meselson and Radding mechanism (figure 5.4(3)).

(2)   The existence of heteroduplex DNA was first inferred from studies

with phage T2. In 1951 **Alfred Hershey and Martha Chase** infected *E. coli* with the wild type (r^+) and a rapid-lysis mutant (r) of T2 and found that when the progeny phage were used to infect a lawn of *E. coli*, about 2 per cent of the plaques were mottled; these plaques had a morphology intermediate between the plaques formed by the wild type and r phages and, unexpectedly, contained approximately equal numbers of r^+ and r phages. Since each mottled plaque resulted from infection by a **single** particle of T2, these phages must have contained both the r^+ and the r alleles. In 1954 **Cyrus Levinthal** showed that most of the phages recovered from the mottled plaques were recombinant for markers flanking the r locus, and he correctly concluded that these partially heterozygous phage genomes were unreplicated intermediates in the recombination process that had arisen by the overlapping of two pieces of DNA of different parental origin, forming heteroduplex molecules which would segregate pure r^+ and r genomes at the next replication (these genomes would have the same structure as the molecules depicted in figure 5.2(6a)).

(3)   In 1968 **Theodore Gurney and Maurice Fox** carried out a remarkably elegant series of transformation experiments using *Streptococcus pneumoniae* and they were able to show that transformation produced segments of heteroduplex DNA, just as predicted by the hybrid DNA models for genetic recombination.

(4)   In 1970 **H. Ch. Spatz and Thomas Trautner** used an ingenious method to demonstrate that bacteria possess a mechanism that can correct mismatched base pairs. Starting with a wild type and a mutant strain of the *Bacillus subtilis* phage SPP1, they separated the light and heavy strands by density-gradient centrifugation (SPP1 has a heavy purine-rich and a light pyrimidine-rich strand of DNA) and annealed the heavy strand from the wild type with the light strand from the mutant and vice versa (figure 5.5). These reciprocal heteroduplex molecules were then used to transfect *B. subtilis*. This is equivalent to transformation but involves the uptake of phage DNA rather than bacterial DNA, and when Spatz and Trautner examined the phage particles released from each transfected bacterium they found that, whereas most of the cells transfected by one species of heteroduplex DNA produced mainly mutant phages, the cells transfected by the reciprocal heteroduplex produced predominantly wild type phages. It is clear that the mispairing of bases within the heteroduplex molecules was usually corrected **before** replication commenced. Furthermore, correction was not a random process, as in one experiment (for example) it was mainly in the direction of the nucleotide sequence carried along the heavy strands.

(5)   Recombination intermediates with the expected chi-form structure were first observed among recombining genomes of phage λ by **Manuel Valenzuela and Ross Inman** in 1975, and by

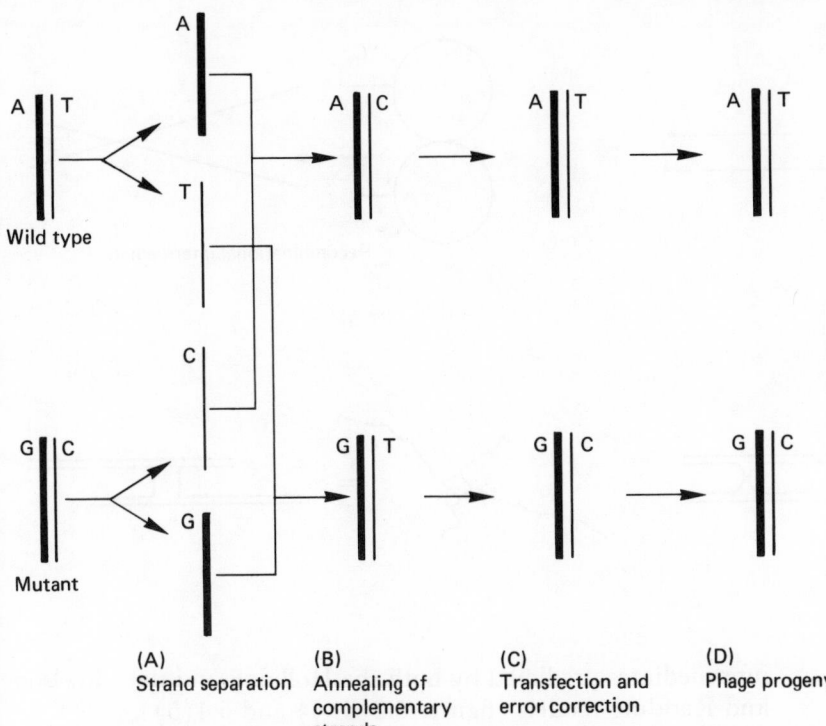

Wild type

Mutant

(A)
Strand separation

(B)
Annealing of
complementary
strands

(C)
Transfection and
error correction

(D)
Phage progeny

Figure 5.5
Error correction during
transfection

Most of the cells transfected with one
species of heteroduplex molecule
produce only wild type progeny phage,
while cells transfected by the
other species of heteroduplex produce
predominantly mutant phages. Thus
error correction seems to precede
replication and, in this experiment,
has been in the direction of the
nucleotide sequences carried by the
heavy strands.

In the figure the mutant is assumed
to have arisen from the wild type by
an A–T to G–C transition mutation.

Huntington Potter and David Dressler using the ColE1
plasmid of *E. coli*. The ColE1 plasmid is a small (6.4 kb) molecule
of circular duplex DNA and normally there are about 20 copies
per cell. **Potter and Dressler** grew *E. coli* ColE1 in the
presence of chloramphenicol, so as to inhibit replication of the
chromosome while allowing continued replication of the plasmid,
until about 1000 copies per cell were present; the cells were then
lysed and the plasmid DNA examined by electron microscopy.
Most of the molecules were plasmid-sized circles but some
were double-sized molecules with a figure-of-eight configuration
(figure 5.6). The extracted DNA was also treated with restriction
endonuclease *Eco*R1, a specific enzyme which cuts each ColE1
circular molecule once at a unique site. If the figure-of-eight
molecules consisted of two interlocking circles, this treatment
would convert them into two separate rod-shaped molecules;
but if they were recombinational intermediates made up of two
molecules of plasmid DNA held together by a cross-over,
they would resolve into chi-shaped structures as, indeed, was
observed. Furthermore, these joint molecules always had two
pairs of arms of equal length; this is precisely the result expected
if the two recombining molecules were held together at regions
of homology. Most of the chi-forms had a simple cross-like
structure but others had arms of double-stranded DNA held
together by a 'box' of single-stranded DNA. These are precisely
the structures of the two isomeric forms of the chi-structure

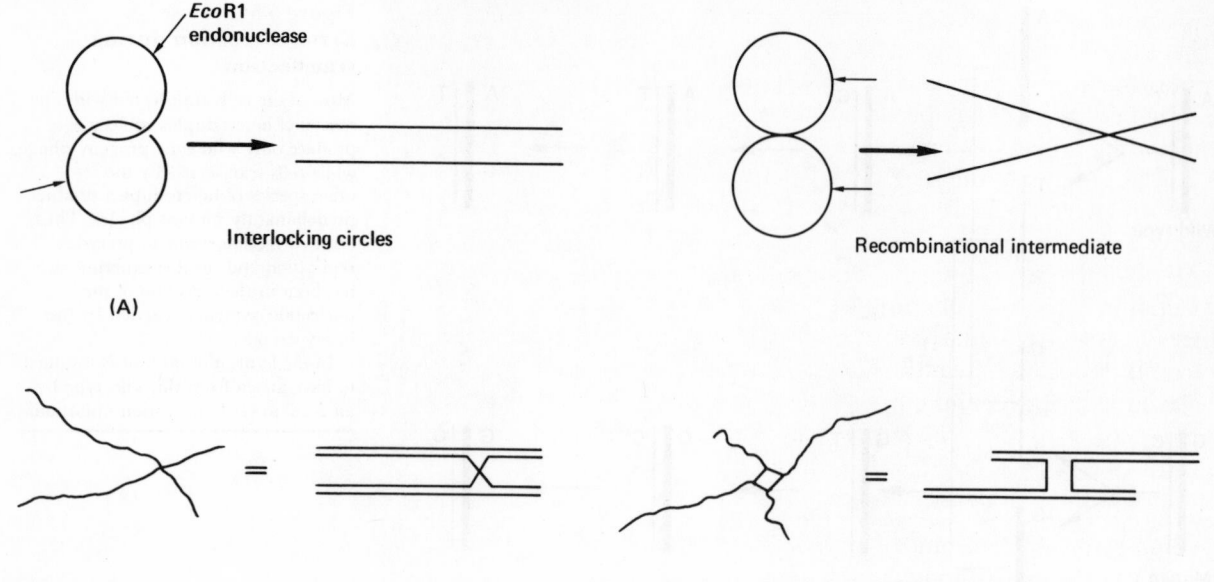

Interlocking circles

(A)

Recombinational intermediate

(B)

**Figure 5.6
Recombinational
intermediates**

(A) If two interlocking circular
molecules are treated with an
endonuclease which cuts each
molecule once at a unique site (small
arrows), the result will be two unit-
length, rod-shaped molecules; but if
the two molecules are held together
by a cross-over, the product will be a
chi-shaped molecule with two pairs of
arms of equal length.
(B) Some of the observed chi-forms.
These are interpreted as the
alternative isomeric forms of the
Holliday intermediate.

intermediates predicted by both the Holliday and the Meselson
and Radding models (figures 5.2(3–4) and 5.4(5)).

More recently, chi-form molecules, also thought to be
recombinational intermediates, have been isolated from yeast
plasmid DNA and from adenovirus infected cells.

(6)   Observations on the RecA protein of *E. coli*, described in the
next section, confirm the prominent role of single-stranded DNA
in the recombinational process.

5.6.1   The RecA protein of *E. coli*

General recombination describes exchanges between homologous or
largely homologous molecules of DNA. It is an enzymatic process and
in *E. coli* there are at least nine genes specifically involved in general
recombination, and mutations within these genes produce strains that
are defective in their ability to carry out general recombination. Most
studied are the mutations within the *recA* gene, which abolish all general
recombination in Hfr × F⁻ crosses and in P1-mediated transductions,
and in the *recB* and *recC* genes which result in a greatly reduced capacity
to recombine.

The *recB* and *recC* genes (abbreviated to *recBC*) together with the
recently discovered *recD* gene, encode the three polypeptides making
up the ATP-dependent DNase known as the *recBCD* DNase or
exonuclease V. This complex enzyme has both exonucleolytic and
endonucleolytic properties and, although its role is not fully understood,
it can bind to duplex DNA and travel along the double helix at about
300 nucleotides per second, creating a loop of unwound DNA; thus
one of its functions appears to be to produce single-stranded DNA. It
can also hydrolyse single-stranded DNA and, in particular, can digest

the strand that is displaced in D-loop formation. Note that *recD* mutants lack the nuclease activities of exonuclease V but, unlike the *recBC* mutants, can still carry out general recombination. The reason for this is not known but it is possible that these mutants retain the DNA-unwinding activity of exonuclease V.

The product of the *recA* gene, first isolated in 1979, is a remarkably versatile protein which, on the one hand, acts as a specific protease to cleave certain repressors (see section 6.7.1) and, on the other hand, is a DNA-dependent ATPase, able to hydrolyse ATP in a DNA-mediated reaction. The role of the ATPase activity has been extensively investigated, particularly in *in vitro* systems where it promotes several recombination-like reactions leading to the formation of D-loops, heteroduplex DNA and chi-form intermediates; these are all reactions predicted by the Meselson and Radding model.

In 1985, **John Flory** demonstrated that the RecA protein promotes the following series of reactions in an *in vitro* system:

1. The RecA protein binds to single-stranded DNA in the presence of ATP; *in vivo* it is thought that single-stranded DNA is produced by the activity of the RecBCD nuclease.
2. The complexed RecA protein rapidly promotes homologous pairing (synapsis) of this single-stranded or partly single-stranded DNA, with homologous sequences on a duplex molecule forming a ternary complex (corresponding to the molecule shown in figure 5.4(2)); in turn, this complex stimulates unwinding of the duplex DNA.
3. The RecA protein more slowly promotes unidirectional strand exchange and D-loop formation; one strand of the duplex molecule is replaced by the single DNA strand derived from the other molecule, thereby creating a joint molecule with a heteroduplex segment (figure 5.4(3) and (4)).

Thus not only does the RecA protein polymerise on to single-stranded DNA and enable the single strand to find its complementary sequence in a molecule of duplex DNA, but it also binds to and unwinds duplex DNA in a reaction that is stimulated by the presence of single-stranded DNA. These findings emphasise the importance of single-stranded DNA in the initiation of general recombination.

It is noteworthy that chi-form molecules, putatively identified as recombinational intermediates, are not found in cells of *E. coli* which lack a functional *recA* gene.

Bacteriophage λ is of interest as in λ-infected cells general recombination can be prompted by either one of two independent systems. Firstly, λ can use the *RecBCD* system of the *E. coli* host cell and, secondly, it can use the phage-encoded Red system. The Red system, encoded by the *redα* and *redβ* genes, produces non-reciprocal recombinants and is the major recombinational pathway, generating about 90 per cent of all phage recombinants. Since the two pathways are independent, λ can undergo general recombination in *recA⁻* hosts although, as would be expected, λ *red⁻* mutants are totally unable to recombine in *recA⁻* host cells.

Very little is known about the Red pathway except that *redα* encodes an endonuclease and *redβ* an associated β-protein; these could be involved in the assimilation of an invading strand by a recipient duplex molecule.

5.6.2　The resolution of chi-form molecules

Although the existence of chi-form molecules and their role in general recombination was first demonstrated in 1975, it is only recently that an enzyme capable of resolving these cruciform structures has been found. Since 1982 **Borries Kemper** and his colleagues have studied a DNase, known as endonuclease VII, encoded by gene 49 of coliphage T4, and have shown that in an *in vitro* system this enzyme can resolve both naturally occurring and artificially synthesised chi-form molecules by introducing a pair of cuts diagonally across the join, exactly as predicted by N–S or by E–W cleavage of the Holliday intermediate (figure 5.2(5)).

Both natural and artificial cruciform molecules could be cleaved in either of these two alternative ways. No special DNA sequences seem to be involved, although the opposite cuts showed a high degree of symmetry so guaranteeing precise resolution of the chi-form intermediate.

Note that endonuclease VII is a phage-encoded enzyme and that no similar enzyme has yet been detected in uninfected cells of *E. coli*; nevertheless, it does establish that such a **resolvase** could exist.

5.7　Chi sites in λ

During the λ lytic cycle, concatenated DNA molecules are obligate intermediates for the encapsidation of phage genomes and they are normally produced by rolling-circle (σ) replication. A feature of this process is that rolling-circle replication can only proceed after the host RecBCD exonuclease V has been inactivated, as otherwise it degrades the tails of the σ-form molecules and prevents replication; this is normally achieved by the product of the λgam^+ gene binding to and inactivating exonuclease V. But if rolling-circle replication is blocked by introducing a *gam⁻* mutation, the only way that concatamers can form is by Red- or RecBC-promoted recombination occurring between the circular monomers produced by θ-type replication; as a result, $\lambda red^- \, gam^-$ cannot form plaques on *recA⁻* bacteria and only forms small plaques on *rec⁺* hosts (this is because the RecBC pathway is relatively inefficient in promoting recombination between λ DNA molecules). In 1975, occasional $\lambda red^- \, gam^-$ variants were found which were able to form large plaques on *rec⁺* hosts and these have been shown to be due to mutations (χ) generating specific nucleotide sequences, called **Chi** (not to be confused with chi-form molecules), which stimulate recombination; these sequences are not present in wild type λ.

These mutations all result from single base pair substitutions and each generates the 8-bp sequence 5′ GCTGGTGG 3′; they have been detected at four widely spaced locations in λ and they have the following properties:

1. They stimulate **only** RecBCD-promoted recombination.
2. The increase in recombination is highest in the immediate vicinity of the Chi site but is still detectable 10 kb to either side of Chi; overall, there is a five-fold increase in the frequency of recombination.
3. The enhancement is always greater to the left of Chi than to the right (as the λ map is drawn in figure 1.11).
4. A Chi site need only be present on one of the recombining molecules.
5. Chi is only effective in promoting recombination when it is correctly oriented in relation to the cohesive-end site (*cos*) of λ; the Chi sequence 5′ GCTGGTGG 3′ must always be on the left strand of λ.

These Chi sites are a natural component of the *E. coli* chromosome (there are about 1000 per genome, or 1 every 4 kb) and there are good reasons to suppose that they play a major part in the recombinational process. They have also been detected in several eucaryotes (including man) and they are conspicuously present in regions of DNA known to undergo a high frequency of recombination or genetic rearrangement.

In 1981 **Gerald Smith and his colleagues** suggested that Chi might act by providing a recognition site for RecBCD enzyme activity, and they proposed a simple model to account for the role of Chi sites in recombination; however, it was not until 1984 that they were able to show that the RecBCD enzyme does recognise Chi and cuts the DNA in its immediate vicinity. The following version of the Meselson and Radding model takes into account these observations (figure 5.7):

1. The RecBCD enzyme attaches to one end of a linear molecule of DNA, such as the end of a fragment of transducing DNA or the end of a piece of DNA introduced during conjugation. This agrees with observations suggesting that the RecBCD enzyme more readily promotes a recombination event when a linear molecule of DNA is involved (figure 5.7(1)).
2. The RecBCD enzyme moves along the DNA, unwinding it as it advances and forming a twin loop; the DNA rewinds behind the advancing enzyme (figure 5.7(2)).
3. When the RecBCD enzyme reaches a correctly oriented Chi site, it cuts the strand carrying the 5′ GCTGGTGG 3′ sequence. This releases the loop on the other strand forming a gap (figure 5.7(3)).
4. As the RecBCD enzyme advances further, rewinding becomes impossible and a single-stranded tail is produced (figure 5.7(4)).
5. This displaced single strand can now be assimilated into the duplex of the other recombining molecule by the activity of the RecA protein (figure 5.7(5)). This structure is identical to the Meselson and Radding intermediate shown in figure 5.4(3).

The final stage in the asymmetric phase of the Meselson and Radding model is branch migration, and this could be promoted by the continued progress of the RecBCD enzyme extending the single-

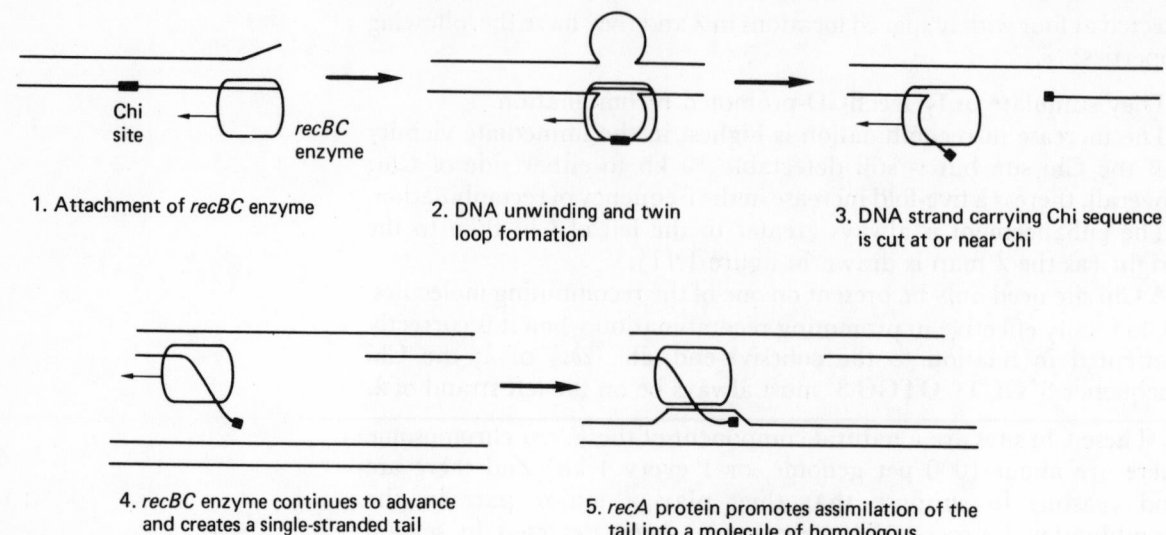

1. Attachment of *recBC* enzyme

2. DNA unwinding and twin loop formation

3. DNA strand carrying Chi sequence is cut at or near Chi

4. *recBC* enzyme continues to advance and creates a single-stranded tail

5. *recA* protein promotes assimilation of the tail into a molecule of homologous double-stranded DNA

Figure 5.7
A model for the role of Chi sites in general recombination

After Smith, G. R., Shultz, D. W., Taylor, A. F. and Triman, K. L., *Stadler Genet. Symp.* **13**, 25 (1981).

stranded tail and by the RecA protein promoting its assimilation into the other recombining molecule.

This model does not explain why, in λ, the Chi sites must be correctly oriented in relation to the cohesive end site. During lytic development the packaging of λ genomes into phage heads can occur **only** if the Ter (terminase) enzyme is present; this enzyme makes staggered nicks at *cos* and cuts off unit length genomes from the concatamer (section 1.5.2). This terminase is also necessary to activate Chi-enhanced recombination. A likely explanation is that Ter enzyme binds asymmetrically to the *cos* site, and after cutting remains temporarily associated with the *cos* sequence on one side of the cut. Thus when Ter binds to and cuts a circular monomer, it will always remain bound to one particular end (the right-hand end) of the now linear molecule (figure 5.8). This means that the RecBCD enzyme can only enter the molecule from the left-hand end and so can only travel along it in one direction; as a consequence the RecBCD enzyme can only recognise Chi when it is in one particular orientation. The cutting at *cos* by the terminase is thought to be a reversible reaction, so the monomers could recircularise after entry of the RecBCD enzyme.

Because of its role in recombination the RecBCD enzyme is frequently referred to as a **recombinase**.

5.8 Recombinational pathways in *E. coli*

The pathways leading to general recombination have been most extensively studied by **Alvin Clark** and his colleagues, largely by isolating and characterising mutants defective in their ability to carry out general recombination and then, from these, isolating further mutants which have regained at least part of their recombinational proficiency.

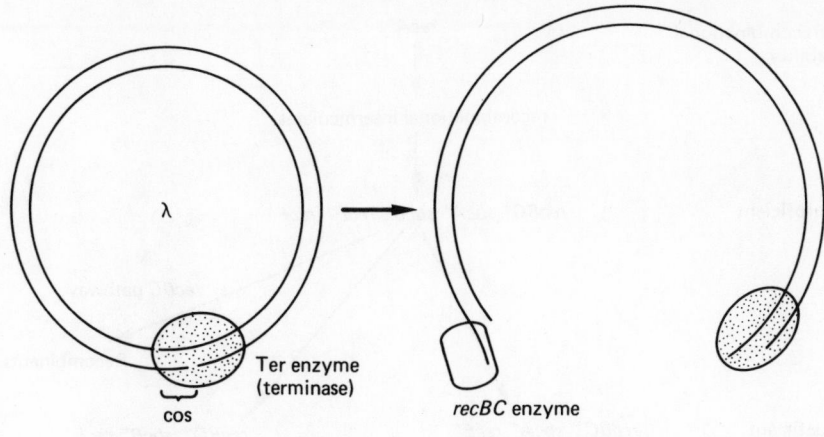

Figure 5.8
The activation of Chi sites in
λ by terminase enzyme

λ

Ter enzyme
(terminase)

cos

recBC enzyme

1. Terminase may bind asymmetrically
   to a *cos* site

2. After cutting, terminase remains bound
   to one particular end of the linear
   molecule so that the *recBC* enzyme can
   only enter the molecule from the other
   end and must always travel along the λ
   DNA in the same direction

In *recA⁻* strains, recombination is almost totally abolished and is reduced some 10^6-fold; it is clear that essentially all homologous recombination requires the active RecA protein to produce recombinational intermediates and that, in wild type cells, these are normally converted into recombinant molecules by the RecBCD protein, exonuclease V. In *recB⁻* or *recC⁻* mutants, however, recombination is not abolished but is only reduced about 100-fold. Some residual recombination still occurs and this is because there are alternative pathways able to resolve the recombinational intermediates; however, in wild type cells these pathways only operate at very low levels, but they can be turned on to maximum activity by the occurrence of further mutations. The best understood are the RecE and RecF pathways, usually studied in strains lacking a functional RecBCD system.

5.8.1   The *RecE* pathway

The *recE⁺* gene encodes exonuclease VIII, which can substitute for exonuclease V in general recombination, but in wild type cells the activity of this gene is repressed by another gene *sbcA⁺*; as a result, *recBC⁻ sbcA⁺* cannot utilise the RecE pathway and are recombination deficient. However, if recombination-proficient reversions of *recBC⁻ sbcA⁺* are isolated, some are *recBC⁻ sbcA⁻*; this is because repression by the *sbcA⁺* gene product is abolished and high levels of exonuclease VIII will be present (*sbc*; suppressor of *recBC* recombination deficiency). In a similar way, mutations to *recE⁻* can be recovered among recombination-deficient derivatives of *recBC⁻ sbcA⁻* (figure 5.9).

The *recE* gene is unusual as it is probably carried on a highly

Figure 5.9
Recombinational pathways in
E. coli

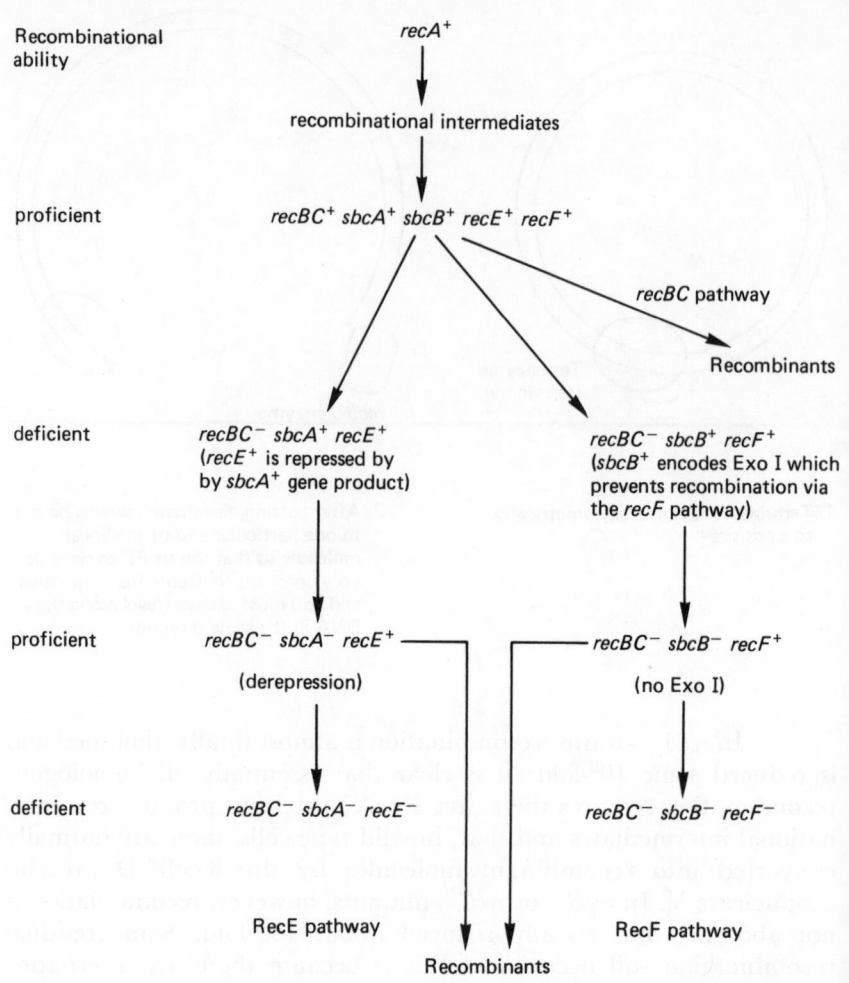

defective prophage genome, lacking an immunity region and normally undetectable in the cell. This **cryptic prophage**, known as Rac, is closely related to λ and, consequently, the RecE exonuclease is very similar to the exonuclease encoded by the λred^+ gene. The $sbcA$ gene is also probably part of the cryptic prophage.

5.8.2 The *RecF* pathway

The *RecF* pathway can also only be detected in $recBC^-$ strains. Another suppressor gene, $sbcB^+$, encodes exonuclease I which prevents recombination proceeding by the RecF pathway, but $recBC^-$ $sbcB^-$ lack this exonuclease and are recombination proficient. The reason for this inhibition is unknown but it is possible that exonuclease I degrades the recombinational intermediates upon which the $recF$ gene product acts.

Exercises

5.1 Assess the evidence for and against: (i) breakage and reunion, (ii) copy choice, and (iii) hybrid DNA models of recombination.

5.2 What is gene conversion and how can it be explained? Why is it important to have linked flanking markers when studying gene conversion?

5.3 In crosses with ascomycetes, occasional asci show 5:3 segregation for a particular mutant site and normal 4:4 segregation for any flanking markers. What two important features of the recombination process does this demonstrate?

5.4 Distinguish between error correction and post-meiotic segregation. How has it been shown that similar phenomena occur in procaryotes?

5.5 Compare and contrast the Holliday and the Meselson & Radding models for recombination. What data cannot be satisfactorily explained by the Holliday model?

5.6 What is a D-loop and how is it formed?

5.7 Do you consider branch migration a necessary component of a recombinational mechanism? Give your reasons.

5.8 How can recombinational intermediates be resolved to produce heteroduplex segments flanked either by recombinant or by parental associations of flanking markers? What is known about this process?

5.9 The figure-of-eight molecules observed by Potter and Dressler (see figure 5.6) are extremely stable but the chi-form molecules produced by *Eco*R1 treatment are rapidly resolved into two linear molecules at room temperature. Explain this difference. Is this further evidence that the chi-forms are recombinational intermediates?

5.10 Outline the role of the RecA protein in general recombination.

5.11 What are Chi sites and what role might they play in the recombinational process?

5.12 Explain why $recA^-$ strains are almost totally recombination deficient whereas $recB^-$ or $recC^-$ strains only have a reduced recombinational proficiency.

References and related reading

Dressler, D. and Potter, H., 'Molecular mechanisms in genetic recombination', *Ann. Rev. Biochem.*, **44**, 45 (1982).

Fogel, S., Mortimer, R. K. and Lusnak, K., 'Mechanisms of meiotic gene conversion or "Wanderings on a foreign strand"', in *The Molecular Biology of the Yeast Saccharomyces, Vol. 1, Life Cycle and Inheritance* (eds J. N. Strathern, E. W. Jones and J. R. Broach), Cold Spring Harbor Laboratories, New York, p. 289 (1981).

Meselson, M. and Radding, C. M., 'A general model for genetic recombination', *Proc. Natl. Acad. Sci. USA*, **72**, 358 (1975).

Radding, C. M., 'Homologous pairing and strand exchange in genetic recombination', *Ann. Rev. Genetics*, **16**, 405 (1982).

Radding, C. M., 'The molecular and enzymatic basis of homologous recombination', in *Bacterial Genetics* (eds J. Scaife, D. Leach and A. Galizzi), Academic Press, London, p. 217 (1985).

Smith, G. R., 'Homologous recombination: the roles of chi sites and RecBC enzyme', in *Bacterial Genetics* (eds J. Scaife, D. Leach and A. Galizzi), Academic Press, London, p. 239 (1985).

Smith, G. R., 'General recombination', in *Lambda II* (eds R. W. Hendrix, J. W. Roberts, F. W. Stahl and R. Weisberg), Cold Spring Harbor Laboratories, New York, p. 175 (1983).

Stahl, F. W., 'Genetic recombination', *Scientific American*, (Feb.), **256**, 53 (1987).

Whitehouse, H. L. K., *Genetic Recombination: Understanding the Mechanisms*, Wiley, New York (1982).

REPAIR AND 6
MUTATION

6.1 DNA repair

At one time it was believed that if DNA was damaged then either the cell would be unable to replicate or it would produce mutant daughter cells by a process corresponding to post-meiotic segregation. It is now abundantly clear that the cell possesses a number of mechanisms, called repair systems, which allow it to remove errors, repair damage and maintain the informational content of its DNA. The ability to repair damaged DNA has probably been significant in evolution, as it enables a species to maintain unchanged its genetic constitution, even in environments that might cause mutations at a high rate. However, it must be remembered that mutation is the ultimate source of all genetic variation, and if repair systems were too efficient the mutation rate could be so reduced that a species could find itself trapped in an evolutionary dead end; presumably, even the efficiency of repair systems must be subject to natural selection.

6.2 Types of DNA damage

DNA is said to be damaged when any change is introduced which is a departure from the normal double-helical structure and normal sequence of nucleotides; some of the principal types of damage are shown in figure 6.1.

(1) **Mismatched bases**. There is an incorrect base on one strand which is unable to H-bond to the corresponding base on the other strand. This may be the result of (a) DNA polymerase selecting the incorrect base for incorporation, which occurs once in every 10^4 incorporations, or (b) the de-amination of cytosine to uracil producing a G–U mismatch or, less frequently, the

Figure 6.1
Types of damage to DNA

Note that with a double-strand break, the two breaks may be exactly opposite each other (upper left) or slightly staggered (upper centre).

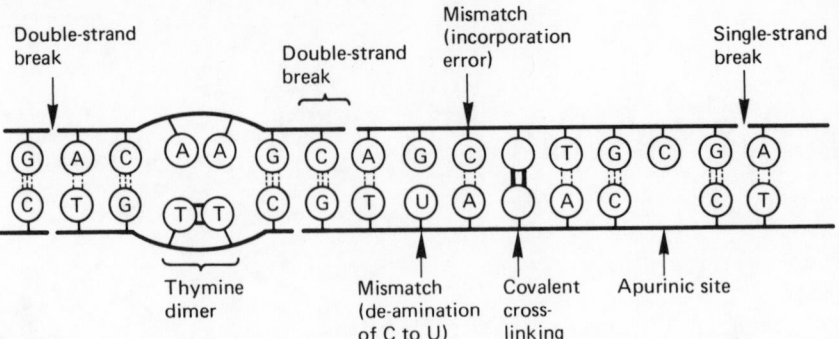

de-amination of adenine to hypoxanthine; these events occur spontaneously but their frequency can be greatly increased by exposure to nitrous oxide.

(2) **Missing bases.** Occasionally purines are lost from DNA by the spontaneous cleavage (by hydrolysis) of the N-glycosylic bond between a purine base and the deoxyribose. This leaves an **apurinic (AP) site** in the DNA. About one purine in 300 per day is lost at 37°C but this rate is accelerated by exposure to alkylating agents; these add an alkyl group to the purine ring (usually guanine), weakening the N-glycosylic bond and increasing the likelihood of depurination.

(3) **Structural alterations.** A wide variety of chemical and physical agents can substantially alter the structure of one or more adjacent bases. The most extensively studied alteration is the **pyrimidine dimer** induced by exposure to ultraviolet irradiation. This causes the formation of covalent bonds between two adjacent pyrimidines on the same strand, distorting the architecture of the double helix, weakening the H-bonds with the bases on the complementary strand and interfering with the forward progress of the replication fork. When DNA polymerase reaches a dimer on the template strand it stalls and, at best, can only resume synthesis at the next priming site on the far side of the dimer; this results in a long single-strand gap, perhaps 1 kb or more, in the newly synthesised strand opposite and on both sides of the dimer. Most commonly these dimers form between two thymines, producing a **thymine dimer**, the so-called model lesion.

These dimers are lethal unless repaired and in a molecule the size of the *E. coli* chromosome (about 4×10^6 bp) a dose of only 100 nJ/mm^2 induces six or seven dimers.

Some antibiotics cause a different type of structural alteration. Mitomycin C, for example, causes the two bases of a complementary pair to form covalent bonds, so preventing the two strands of DNA from separating during replication.

(4) **Broken phosphodiester bonds.** Many agents, particularly peroxides, ionising radiations and endonucleases, cause breaks in

one or both strands of the DNA backbone (single-strand and double-strand breaks respectively).

6.3 DNA repair systems

Each of the foregoing types of damage may be repaired by at least one of the remarkably diverse **repair systems**. There are two general types of repair system; firstly, systems which **reverse** the damage and, secondly, systems which **excise** the damage and replace it with an undamaged sequence retrieved by templating against the intact complementary strand. These types of repair are best illustrated by two well-understood mechanisms, **photoreactivation** and **excision repair**, largely confined to non-replicating regions of the genome. Although this pre-replicational repair is clearly the most efficient way to repair damage there is another group of systems, usually but incorrectly referred to as repair mechanisms, which **tolerate** rather than repair the damage. These permit replication to **by-pass** damage on the template strand and to maintain the continuity of the genome; these include **post-replicational repair** and **transdimer synthesis**. It is interesting to note that DNA is the only macromolecule that can be repaired by cellular mechanisms in this way.

We are principally concerned with systems that repair the model lesion, the thymine dimer, but in passing we will mention some of the other systems which cannot repair dimers but which can repair other types of damage.

6.4 The reversal of damage to DNA — photoreactivation

Photoreactivation (PR) was first observed by **Albert Kelner** in 1949; he found that the number of actinomycetes surviving a large dose of ultraviolet irradiation could be increased several hundred fold by exposing the irradiated bacteria to an intense source of visible light. Photoreactivation, as this was later called, is one of the most striking examples of a repair system and an important mechanism for the repair of pyrimidine dimers.

Photoreactivation is a three-step process: (i) a PR enzyme, DNA photolyase, recognises the distortion to the double helix caused by the dimer and binds to it; (ii) the DNA–enzyme complex absorbs visible light and, using this light energy, cleaves the dimer into its two component pyrimidines; and (iii) the enzyme is released, restoring the native DNA (figure 6.2).

Note that photoreactivation does not require the presence of an intact complementary DNA strand, no DNA synthesis is involved and, because visible light is required to activate the DNA–enzyme complex, it cannot take place in the dark.

In *E. coli* DNA photolyase is encoded by two genes, *phrA* and *phrB*;

Figure 6.2
Photoreactivation

No DNA synthesis is involved and the
dimers are enzymatically uncoupled.

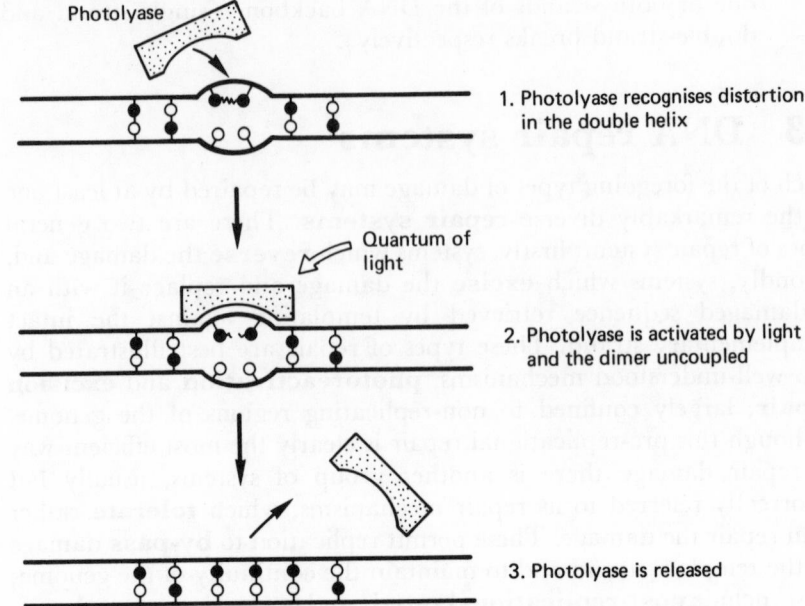

Photolyase

1. Photolyase recognises distortion
   in the double helix

Quantum of
light

2. Photolyase is activated by light
   and the dimer uncoupled

3. Photolyase is released

similar enzymes have been isolated from many different bacteria, plants
and animals, including man.

6.5 Excision repair

Excision repair, discovered by **Robert Setlow** in 1964, is a very efficient
and very accurate system which repairs DNA by cutting out mismatched
or damaged bases and replacing them with a tract of new and intact
DNA. Since light plays no part in the process, irradiated cells
repair their DNA and recover their ability to replicate even when kept
in the dark. Not only can *E. coli* use excision repair to repair its own
DNA but it can also repair the DNA of certain irradiated phages (such
as T1 and T3) which are unable to repair their own DNA; this is
known as **host cell reactivation**.

The repair of a pyrimidine dimer normally involves the following
four enzymatic steps (figure 6.3, left-hand pathway):

1. **Incision.** A repair or incision endonuclease, in this instance the
   uvrABC endonuclease, recognises the distortion caused by the dimer
   (different endonucleases may recognise other types of damage) and
   cuts a phosphodiester bond on the 5' side of the dimer.

   The next two steps are carried out concurrently by DNA
   polymerase I.

2. **Excision.** DP I uses its 5' to 3' exonuclease activity to remove, or
   excise, a tract of DNA between 7 and 20 nucleotides long and
   including the dimer.

3. **Polymerisation.** It uses its 5' to 3' polymerase activity to fill the
   gap using the undamaged daughter strand as a template. It

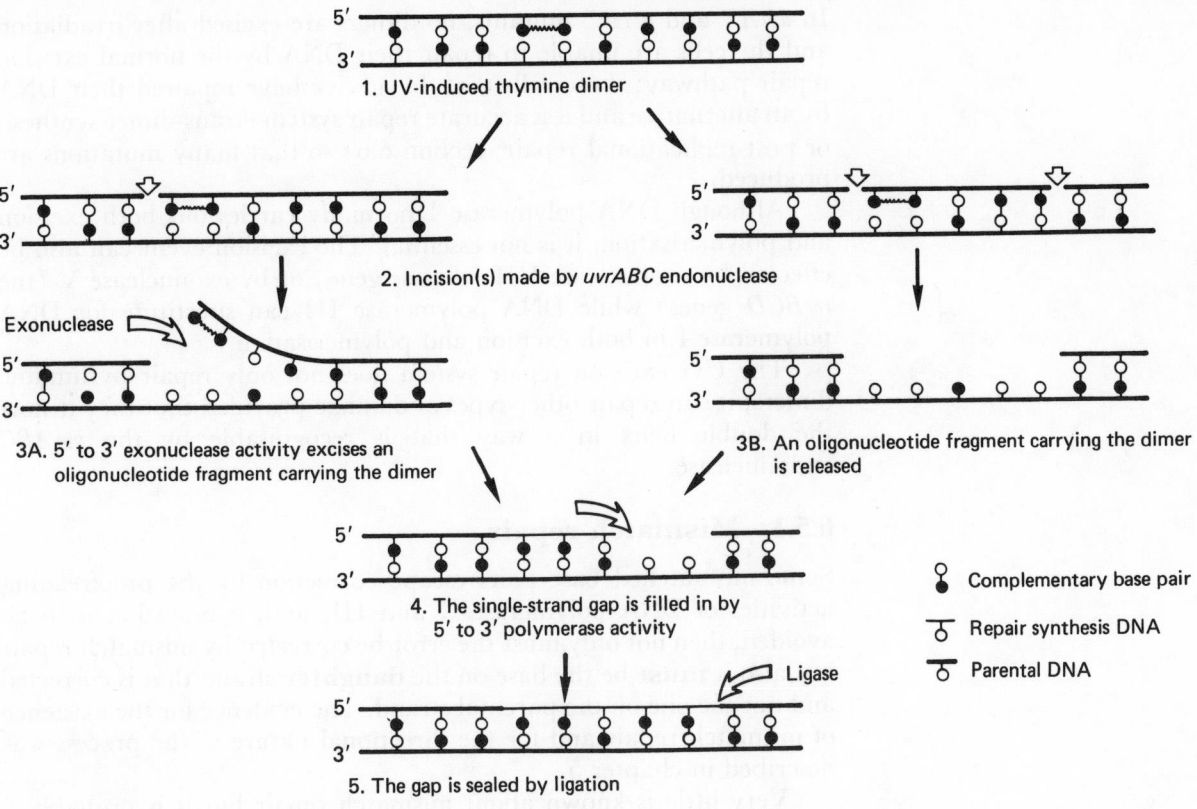

1. UV-induced thymine dimer

2. Incision(s) made by *uvrABC* endonuclease

Exonuclease

3A. 5′ to 3′ exonuclease activity excises an oligonucleotide fragment carrying the dimer

3B. An oligonucleotide fragment carrying the dimer is released

4. The single-strand gap is filled in by 5′ to 3′ polymerase activity

Ligase

5. The gap is sealed by ligation

Complementary base pair

Repair synthesis DNA

Parental DNA

Figure 6.3
Excision repair

At the left is shown the normal pathway for excision repair. The pathway shown at the right has, so far, only been detected *in vitro* and involves the *uvrABC* endonuclease making two incisions on the same strand so that the damaged tract can be excised without the need for exonuclease activity.

recognises the exposed 3′–OH end and synthesises a new strand while, at the same time, displacing the old one.

4. **Ligation.** Finally, exonuclease cuts off the damaged segment and the remaining single-strand break is sealed by polynucleotide ligase.

Quite recently, studies using an *in vitro* system have shown that the *uvrABC* endonuclease can also make two incisions about 12 nucleotides apart, one on each side of the dimer, allowing excision of the damaged tract without the need for exonuclease activity (figure 6.3, right-hand pathway). However, it is not yet known whether this mechanism operates *in vivo*.

This is the major excision repair system in *E. coli* and it is sometimes called **short-patch repair** because the average length of the excised tract is only about 20 nucleotides or less.

Four widely spaced genes are involved in dimer excision, *uvrA*, *uvrB*, *uvrC* and *uvrD* (*uvr* for *ultraviolet* repair). *uvrA*$^+$ and *uvrB*$^+$ encode the two major components of the *uvrABC* endonuclease while the *uvrC*$^+$ protein, also a component of the endonuclease, is required for maximum activity; the function of the protein encoded by *uvrD*$^+$ is not known.

These genes were first identified by isolating mutants which had both greatly increased sensitivity to ultraviolet irradiation and increased frequencies of ultraviolet-induced mutation; this is because the *uvr*$^-$ mutants either lack or have reduced amounts of the incision endonuclease.

In $uvrA^-$ and $uvrB^-$ mutants, no dimers are excised after irradiation and the cells are unable to repair their DNA by the normal excision repair pathway; those cells that do survive have repaired their DNA by an alternative and less accurate repair system (trans-dimer synthesis or post-replicational repair; section 6.6) so that many mutations are produced.

Although DNA polymerase I normally carries out both excision and polymerisation, it is not essential. The excision event can also be effected by exonuclease VII (the *xse* gene) or by exonuclease V (the *recBCD* genes) while DNA polymerase III can substitute for DNA polymerase I in both excision and polymerisation.

The Uvr excision repair system does not only repair pyrimidine dimers; it can repair other types of damage provided that they distort the double helix in a way that is recognisable by the *uvrABC* endonuclease.

6.5.1 Mismatch repair

Some mismatched base pairs escape correction by the proofreading activities of DNA polymerases I and III, and, if mutation is to be avoided, then not only must the error be corrected by mismatch repair but also it **must** be the base on the **daughter** strand that is corrected and not the one on the parental strand. The evidence for the existence of mismatch repair and for the directional nature of the process was described in chapter 5.

Very little is known about mismatch repair but it is probably a form of excision repair, which nucleotide of the mismatch is replaced being determined by the degree of methylation of the DNA (box 6.1). After replication the DNA is only hemimethylated and no methylated bases are present on the newly synthesised daughter strand. It seems likely that the mismatch repair system recognises the undermethylated daughter strand in the interval between replication and methylation, and corrects the mismatch by removing the misincorporated base on the daughter strand.

In support of this is the observation that dam^- mutants (lacking adenine methylase) are not only more sensitive to ultraviolet irradiation but they have increased rates of mutation (that is, dam^- acts as a **mutator** gene); this suggests that under-methylation in **both** strands deprives the cell of the ability to discriminate between the parental and daughter strands.

6.5.2 N-glycosylase excision repair systems

The cell contains a number of repair enzymes called **N-glycosylases** (or N-glycosidases) which excise as free bases certain types of incorrect or damaged bases. The best known is uracil-N-glycosylase which cleaves uracil from either single-stranded or double-stranded DNA leaving an **apyrimidinic (AP)** site.

Although uracil is normally found only in RNA (where it replaces thymine) dUTP is not infrequently misincorporated into DNA in place

---

Box 6.1 The modification of DNA by methylation

In recent years it has become clear that bacterial DNA is usually methylated at certain target sites, usually 4 to 6 bp symmetrical sequences; such sequences occur, on the average, once in every 256–4096 nucleotides and contain either 6-methyladenine or 5-methylcytosine. In *E. coli*, the enzymes DNA adenine methylase (encoded by the *dam* gene) and DNA cytosine methylase (*dcm*) recognise the symmetrical sequences 5'GATC3' and 5'CC$\frac{A}{T}$GG3' and methylate either the adenine or cytosine bases; because the sequences are symmetric, both strands will be methylated within the target site:

$$
\begin{array}{cc}
\overset{*}{} & \overset{*}{} \\
G\,A\,T\,C & C\,C\,A\,G\,G \\
C\,T\,A\,G & G\,G\,T\,C\,C \\
\underset{*}{} & \underset{*}{}
\end{array}
$$

When the DNA replicates, the methyl groups on the parental strands are retained but the new daughter strand is unmethylated; the methyl groups are added subsequently by methylase activity acting on the target sequences; thus:

hemimethylated   fully methylated

In addition, nearly all bacterial cells have a **restriction modification system** designed to protect the host cell DNA and to break down any foreign DNA (particularly the DNA of an infecting phage) that might enter the cell. Most species or strains have a specific **restriction endonuclease** which cleaves DNA within or besides every copy of a particular target sequence. At the same time, each strain also has a specific methylase which methylates the DNA within the same target sequence and protects it from restriction endonuclease attack. Because different species and strains have different restriction modification systems, which recognise different target sequences, the host cell and foreign DNA will be methylated at different sites; thus foreign DNA introduced into *E. coli* will be recognised by the host cell restriction endonuclease and degraded.

---

of dTTP, so forming an adenine–uracil base pair; this probably occurs once in every 1200 incorporations. Another source of uracil in DNA is the spontaneous *in situ* de-amination of cytosine to form a guanine– uracil base pair. These mismatches cannot be distinguished from normal A–T and G–C base pairs by either DNA polymerase I or III, and so escape proofreading, and since they do not produce a distortion in the double helix they cannot be corrected by the *uvrABC* system of excision repair. Since the uncorrected mismatch would be mutagenic, there is a specific mechanism for replacing these unwanted uracils in DNA (figure 6.4); this type of repair directly recognises the abnormal base and so avoids the necessity for distinguishing between the parental and daughter strands.

The uracil in the DNA is recognised by a uracil-specific enzyme,

Figure 6.4
Uracil–N–glycosylase
excision repair

1. DNA with a misincorporated uracil

Uracil-N-glycosylase

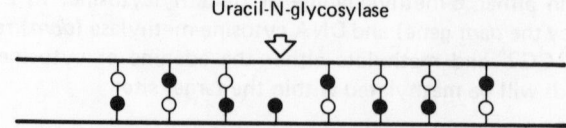

2. Uracil-N-glycosylase excises the uracil base

AP endonuclease

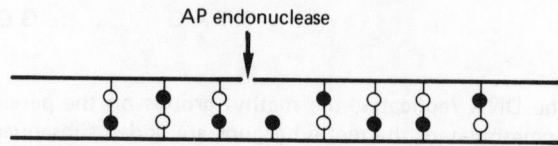

3. AP endonuclease makes an incision adjacent to the defect

DP I

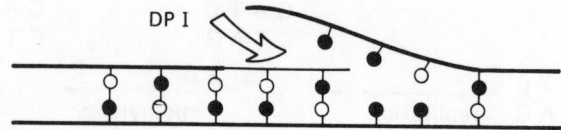

4. DP I uses its 5′ to 3′ exonuclease and polymerase activity
   to replace a tract including the AP site

uracil-N-glycosylase (encoded by the *ung* gene), which cleaves the N-glycosylic bond between the uracil and its deoxyribose, releasing a free uracil and leaving an apyrimidinic site in the DNA. This apyrimidinic site is recognised by an **AP endonuclease** which cuts the strand on the 5′ side of the missing base. The repair is completed by DNA polymerase I using its 5′ to 3′ exonuclease and polymerase activities to excise a piece of DNA including the apyrimidinic site, and to replace it with a correct sequence templated against the parental strand.

Other N-glycosylases are able to recognise and excise other damaged or abnormal bases, such as hypoxanthine and 3-methyladenine.

6.6   Post-replicational repair (PRR)

Since the early 1960s it was recognised that repair and recombination share a similar set of enzymatic activities — endonuclease nicking, exonuclease digestion, polymerisation and ligation — and this led to the prediction that strains might be found which, on the one hand, would be unable to carry out genetic recombination and, on the other hand, be radiation sensitive because of an inability to carry out one or other of the repair processes; the *recA⁻*, *recB⁻* and *recC⁻* strains of *E. coli*, first isolated in 1965 (section 5.6.1), have just these properties.

In 1968, **Paul Howard-Flanders** found that different strains had different sensitivities to ultraviolet irradiation; whereas *uvrA⁻* (excision

repair defective) and *recA*⁻ (recombination deficient) were more
sensitive to ultraviolet irradiation than the wild type, the *uvrA*⁻ *recA*⁻
double mutant was extremely sensitive and could only tolerate about
1 dimer per genome (table 6.1). Furthermore, when an irradiated
recA⁻ strain was stored in the dark under conditions which did not
permit replication, the cells slowly recovered their ability to replicate
and to survive; on the other hand, storage of a *uvrA*⁻ mutant did not
increase the survival. This is because in *uvrA*⁺ strains (wild type
or *recA*⁻) most of the thymine dimers are removed by the efficient and
accurate excision repair system which functions independently of
replication; however, since the excision defective *uvrA*⁻ strain could
tolerate as many as 50 dimers per genome, it was clear that there was
a further repair system functioning during or after replication and
dependent upon the presence of the *recA*⁺ gene. This is **post-
replicational repair** and it is the only system that can repair (or,
strictly speaking, tolerate) DNA that is damaged in both strands at
the same site.

The ability of a cell to carry out DNA repair can have a significant
effect on the frequency of induced mutation. **Evelyn Witkin** has
extensively investigated ultraviolet-induced mutagenesis in *E. coli* and
in one set of experiments both *uvrA*⁺ and *uvrA*⁻ strains were given a
dose of 2 mJ/mm², sufficient to induce about 120 dimers per genome.
In the *uvrA*⁺ cells, most of the dimers were corrected and the mutation
frequency to streptomycin resistance was only 0.4 mutations per 10^7
survivors. However, in *uvrA*⁻ cells, which lack the excision repair
system, the mutation frequency was 200 per 10^7 bacterial survivors.
Not only must the excision repair process be extremely accurate but
most ultraviolet-induced mutations were the result of the continued
presence of dimers in the *uvrA*⁻ strain. Witkin calculated that the
excision repair process makes less than one mistake for every 10^6 dimers
excised.

Witkin also found that ultraviolet irradiation cannot induce
mutations in *recA*⁻ strains. This suggested that ultraviolet-induced
mutations were induced **after** replication and during the post-
replicational repair of any dimers that had not been corrected by the
excision repair system; thus post-replicational repair is an **error-prone**
process and makes more mistakes than the error-free excision repair
system — more mismatched bases are inserted during the repair process
and these subsequently lead to mutation.

Genotype	Phenotype	UV dose ($\mu J/mm^2$)	Dimers/ genome
uvrA⁺ *recA*⁺	wild type	50	3200
uvrA⁻ *recA*⁺	excision-repair deficient	0.8	50
uvrA⁺ *recA*⁻	recombination deficient	0.3	20
uvrA⁻ *recA*⁻	excision-repair deficient *and* recombination deficient	0.02	1.3

**Table 6.1
Doses of ultraviolet
irradiation for 37 per cent
survival of *E. coli* K12**

Data from Howard-Flanders, P.,
Ann. Rev. Biochem. **37**, 175 (1968).

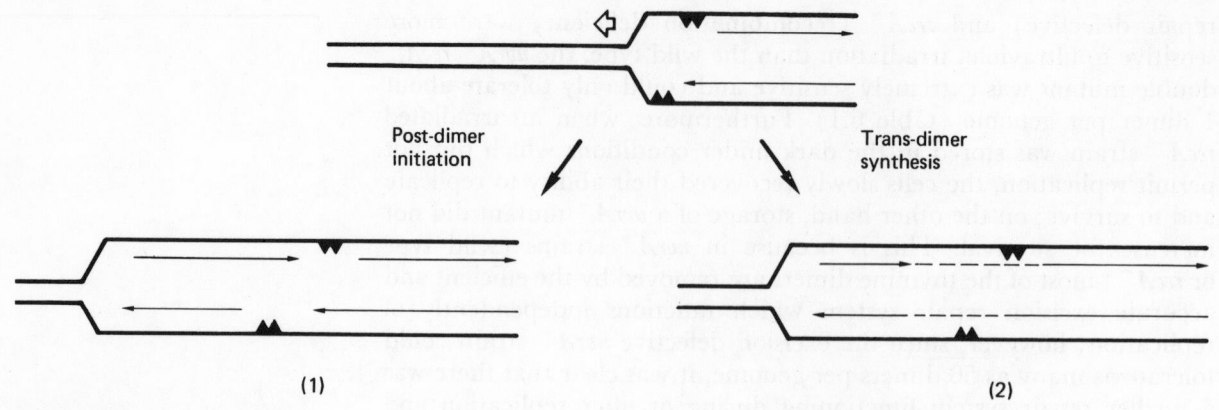

Post-dimer
initiation

Trans-dimer
synthesis

(1)

(2)

Figure 6.5
By-passing unexcised dimers

With post-dimer initiation, replication recommences just beyond the dimer leaving a gap in the daughter strand. In trans-dimer synthesis, nucleotides are inserted opposite the dimer but frequent mismatches are introduced on both sides of the dimer.

6.6.1   By-passing unexcised dimers

When DNA replicates, an unexcised dimer cannot be used as a template by DNA polymerase. As a result, replication is temporarily stalled although it recommences a few seconds later from a site some distance beyond the dimer, leaving a gap opposite the unexcised dimer (figure 6.5(1)); this is known as **post-dimer initiation** but the mechanism is not understood.

An alternative mechanism for by-passing unexcised dimers is **trans-dimer synthesis** (figure 6.5(2)); this is part of the SOS system (section 6.7) and, although it inserts nucleotides opposite a dimer, it produces frequent errors of incorporation.

6.6.2   A model for post-replicational repair

Following post-dimer initiation, both strands are damaged at the same site — one has a gap and the other a dimer — and the only way this damage can be tolerated is by a recombination event which replaces the DNA missing from the gap opposite the dimer in the first daughter molecule with a tract of homologous single-stranded DNA derived from the second daughter duplex. Every unexcised dimer must be by-passed in this way and the process can **only** occur in cells having a functional system of general recombination. This type of 'repair', unlike excision repair, occurs after replication and is usually called **post-replicational repair** (PRR).

The molecular basis of PRR is not established but a plausible scheme based on the Meselson and Radding model for general recombination (section 5.5) is shown in figure 6.6.

It is important to note that PRR does not remove dimers. The dimers are still present in the 'repaired' molecule but they are now opposite an intact nucleotide sequence and can be repaired by excision repair; if they are not repaired then the PRR process must be repeated after every replication cycle. The importance of PRR is that it permits the replication and survival of molecules of DNA that otherwise would be unable to replicate.

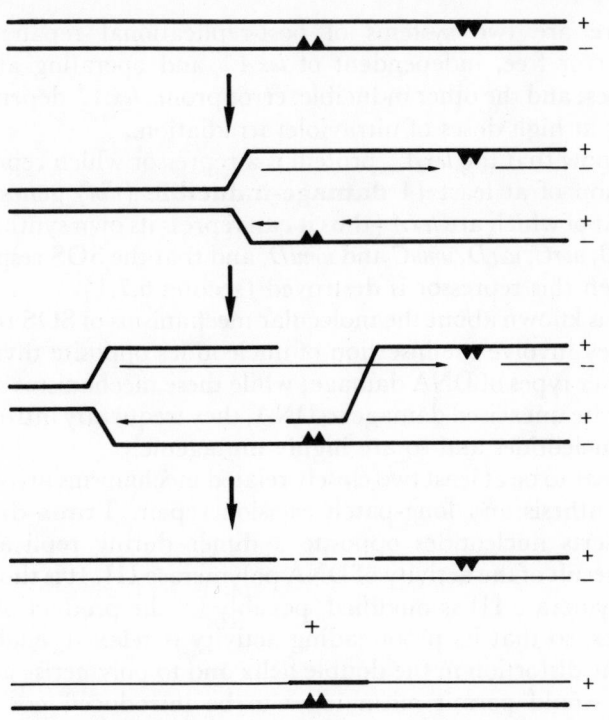

1. Unexcised dimers are present in a molecule of parental DNA

2. At the next replication, post-dimer initiation leaves gaps opposite each dimer

3. The undamaged parental strand on the upper daughter molecule is nicked. The strand dissociates and is assimilated into the gap in the newly synthesised strand of the lower daughter molecule

4. The gap in the upper daughter molecule is filled in by repair synthesis (exonuclease, polymerase and ligase activity)

Figure 6.6
Post-replicational repair

The gaps left by post-dimer initiation are repaired by a recombinational mechanism; this retrieves the missing information from the corresponding undamaged strand in the sister molecule.

6.7 The SOS hypothesis

The proteins required for excision repair and post-replicational repair are present in the cells of *E. coli* at all times but only in small quantities, just enough to repair the occasional damage that occurs under normal conditions (the *uvrA, uvrB, uvrC* and *recA* proteins) and to promote general recombination (the *recA* protein); these low-level **constitutive** systems are very accurate, or **error free**, and so cause very few mutations. However, there is another group of repair systems which is **induced** by the presence of an extensive amount of damaged DNA and which are emergency (SOS) systems for repairing DNA. These systems introduce many mismatches during repair and so are highly mutagenic or **error prone**.

The SOS hypothesis was proposed by **Miloslav Radman** in 1974 following two important discoveries. The first was that the error-prone system of repair is totally dependent on the presence not only of $recA^+$ but also of another gene, $lexA^+$; $lexA^-$ strains are able to carry out general recombination and, although they can carry out post-replicational repair, the ultraviolet-induced mutability is totally abolished. Thus not only is $lexA^+$ required for error-prone repair but $lexA^-$ strains must be able to repair lesions by an error-free system of post-replicational repair. The second discovery was that low doses of ultraviolet irradiation ($500\ nJ/mm^2$ or less) which induce fewer than 30 dimers per genome do **not** result in ultraviolet-induced mutations whereas higher doses do; clearly, error-prone repair only functions after a relatively high dose of ultraviolet irradiation.

Thus there are two systems of post-replicational repair: one constitutive, error free, independent of *lexA*[+] and operating at low ultraviolet doses; and the other inducible, error prone, *lexA*[+] dependent and operating at high doses of ultraviolet irradiation.

We now know that the *lexA*[+] protein is a repressor which represses the transcription of at least 14 **damage-inducible** (*din*) genes, the most important of which are *lexA* (thus it can repress its own synthesis), *recA, uvrA, uvrB, uvrC, uvrD, umuC* and *umuD*, and that the SOS response is induced when this repressor is destroyed (section 6.7.1).

Very little is known about the molecular mechanisms of SOS repair except that they involve the insertion of nucleotides opposite thymine dimers and other types of DNA damage; while these mechanisms allow the cell to tolerate unexcised damage to DNA, they frequently introduce mismatched nucleotides and so are highly mutagenic.

There appear to be at least two closely related mechanisms involved, trans-dimer synthesis and long-patch excision repair. **Trans-dimer synthesis** inserts nucleotides opposite a dimer during replication, probably as a result of the activity of DNA polymerase III. It is thought that DNA polymerase III is modified, possibly by the product of one of the *din* genes, so that its proofreading activity is relaxed, enabling it to tolerate the distortion in the double helix and to polymerise across a dimer; this would permit mismatches to be introduced not only opposite the dimer but on either side of it. **Long-patch excision repair** resembles normal (short-patch) excision repair except that the excised tract is typically about 1500 nucleotides long and the gap-filling activity is carried out by an error-prone DNA polymerase (DP III?)

The role of the *recA* protein in SOS repair is unknown, but because it binds so readily to single-stranded DNA it might well protect the single-stranded region containing the dimer from nucleolytic attack until after repair synthesis.

The several pathways for repair in *E. coli* are summarised in figure 6.7.

6.7.1 A model for SOS induction

In the absence of DNA damage, both the *lexA* repressor and the *recA* protein are produced constitutively at low levels and the *lexA* repressor binds to identical 20 bp long operator sequences present within the regions controlling the expression of each *din* gene; as a consequence, all the SOS-inducible loci are repressed and only low-level constitutive synthesis can occur (figure 6.8(A)).

In the presence of ultraviolet-damaged DNA (figure 6.8(B)), all the *din* genes are coordinately induced by the damage activating the specific endopeptidase activity of the *recA*[+] protein (section 5.6.1). It is thought that small molecules released from the damaged DNA, possibly either oligonucleotides or single-stranded fragments of DNA, act as effector molecules, bind to the *recA*[+] protein and activate its protease activity. Using this activity, the *recA*[+] protein recognises and destroys the *lexA*[+] repressor so that all the *din* genes, including *lexA*[+] and *recA*[+], are fully induced. Once the damage is repaired there will

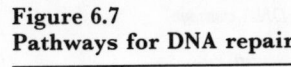

Figure 6.7
Pathways for DNA repair

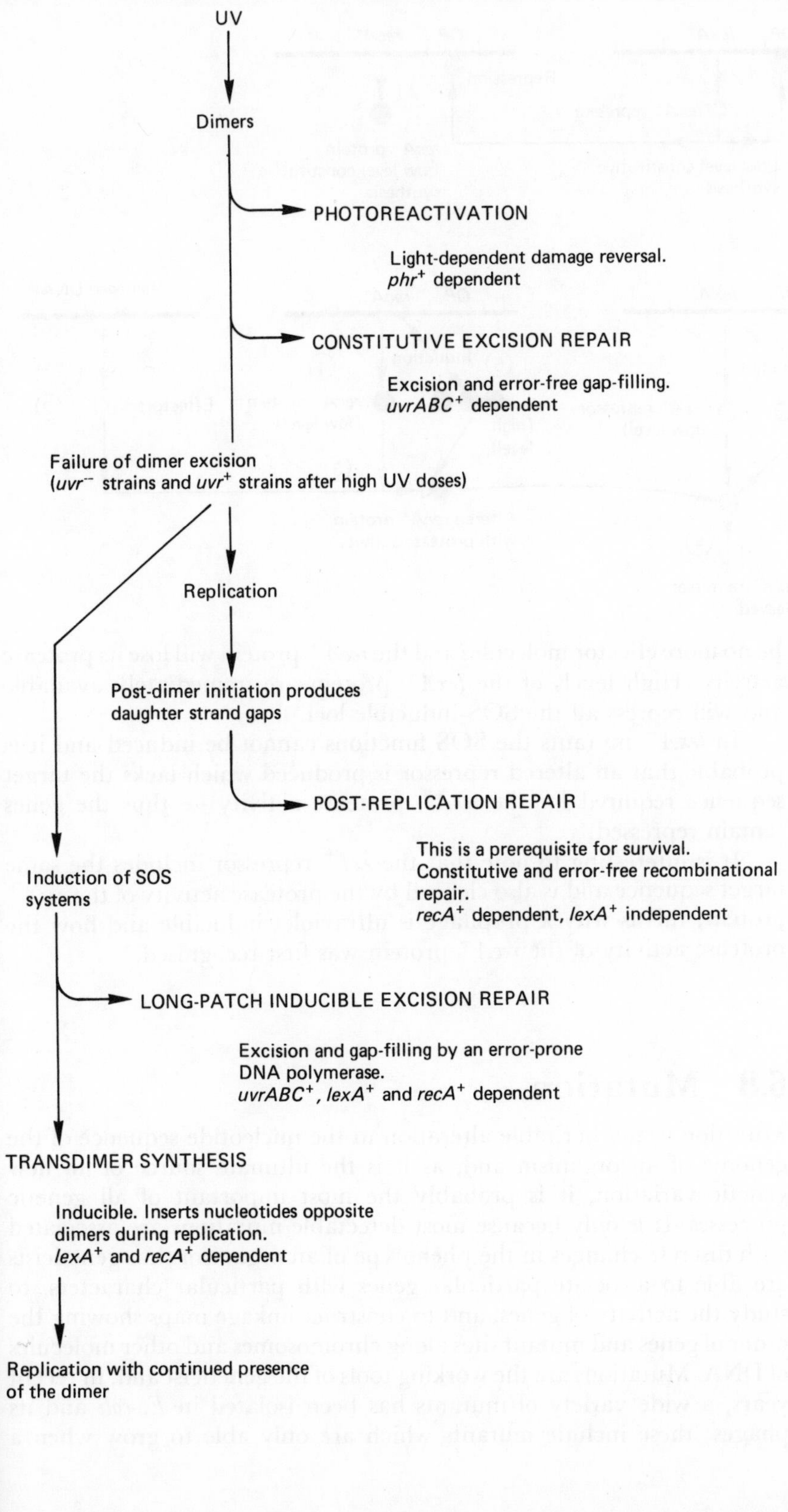

UV

↓

Dimers

→ PHOTOREACTIVATION

Light-dependent damage reversal.
phr^+ dependent

→ CONSTITUTIVE EXCISION REPAIR

Excision and error-free gap-filling.
$uvrABC^+$ dependent

Failure of dimer excision
(uvr^- strains and uvr^+ strains after high UV doses)

↓

Replication

Post-dimer initiation produces
daughter strand gaps

→ POST-REPLICATION REPAIR

This is a prerequisite for survival.
Constitutive and error-free recombinational
repair.
$recA^+$ dependent, $lexA^+$ independent

Induction of SOS
systems

→ LONG-PATCH INDUCIBLE EXCISION REPAIR

Excision and gap-filling by an error-prone
DNA polymerase.
$uvrABC^+$, $lexA^+$ and $recA^+$ dependent

TRANSDIMER SYNTHESIS

Inducible. Inserts nucleotides opposite
dimers during replication.
$lexA^+$ and $recA^+$ dependent

↓

Replication with continued presence
of the dimer

(A) *No DNA damage*

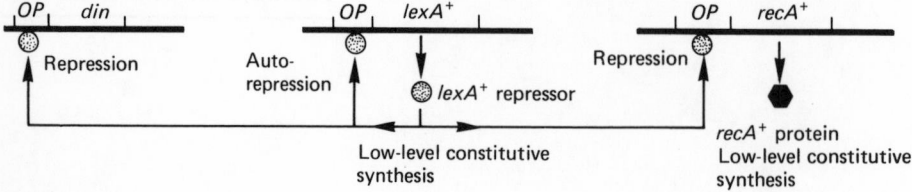

(B) *DNA damaged*

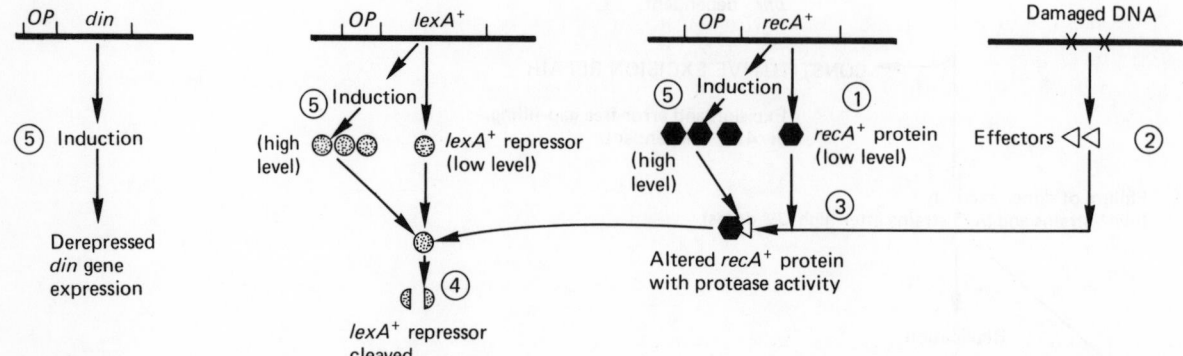

Figure 6.8
A model for SOS induction

(A) In the absence of DNA damage the *din*, *lexA*⁺ and *recA*⁺ genes are repressed by the *lexA*⁺ repressor.
(B) The *recA*⁺ protein is normally produced at a low level (1). After DNA damage effector molecules (2) bind to the *recA*⁺ protein and stimulate its protease activity (3); as a result the *lexA*⁺ repressor is cleaved and inactivated (4) and the *din*, *lexA*⁺ and *recA*⁺ genes are co-ordinately induced (5).

be no more effector molecules and the *recA*⁺ protein will lose its protease activity. High levels of the *lexA*⁺ protein are immediately available and will repress all the SOS-inducible loci.

In *lexA*⁻ mutants the SOS functions cannot be induced and it is probable that an altered repressor is produced which lacks the target sequence required for the *recA*⁺ protease activity — thus the genes remain repressed.

It is interesting to note that the *λcI*⁺ repressor includes the same target sequence and is also cleaved by the protease activity of the *recA*⁺ protein; this is why λ prophage is ultraviolet inducible and how the protease activity of the *recA*⁺ protein was first recognised.

6.8 Mutation

Mutation is any heritable alteration in the nucleotide sequence of the genome of an organism and, as it is the ultimate source of all new genetic variation, it is probably the most important of all genetic processes. It is only because most detectable mutations are associated with discrete changes in the phenotype of an organism that geneticists are able to associate particular genes with particular characters, to study the activity of genes, and to construct linkage maps showing the order of genes and mutant sites along chromosomes and other molecules of DNA. Mutations are the working tools of the geneticist and, in recent years, a wide variety of mutants has been isolated in *E. coli* and its phages; these include mutants which are only able to grow when a specific growth factor is provided (**auxotrophs**), mutants unable to ferment particular carbon sources, mutants resistant to a wide range of drugs, and mutants unable to carry out one or other of the basic cellular processes such as replication, recombination or repair.

These mutations from the wild type to mutant are referred to as **forward-mutations**, whereas those from mutant back to the wild type are known as **back-mutations** or **reverse-mutations**.

6.8.1 The isolation of mutants

Most genetic studies with *E. coli*, such as those described in this book, start with the isolation of mutants. In *E. coli* (as in most procaryotes) this is a relatively straightforward procedure as it is a haploid organism and any mutation resulting in altered gene expression may have an immediate and detectable effect on the phenotype. The investigator usually requires to isolate mutants with a particular phenotype and, since mutation is a rare and random event, special methods may be used to facilitate their detection. Note that it is not usually possible to isolate particular mutant genotypes as often several structural genes, together with their regulatory elements, may be involved in the production of a particular phenotype.

Selection can be used for directly isolating certain types of forward-mutation, notably mutations to **drug or phage resistance** and the wild type bacterial strain is plated on a medium on which only the desired mutant phenotype can grow. For example, mutations to chloramphenicol resistance can be isolated by plating 10^7-10^8 *E. coli* cells on medium containing chloramphenicol; the few colonies that grow will be chloramphenicol resistant.

Screening methods are of more general use and they enable large numbers of cells to be tested rapidly for the required phenotype, usually after treatment with a mutagen (section 6.9). One screening technique commonly used to isolate **non-fermenting mutants** uses an indicator medium. For example, lactose-negative (Lac$^-$) mutants can be distinguished from among several thousand wild-type cells by plating on a fully nutrient medium containing lactose and the colourless indicator dye 2,3,5-triphenyltetrazolium chloride. At the high pH produced within Lac$^-$ colonies, the tetrazolium is reduced to dark red formazan whereas in the Lac$^+$ colonies the pH is lowered by the fermentation of lactose and the reaction is inhibited; as a result, the Lac$^+$ colonies are whitish and the Lac$^-$ mutants are a dark blood-red.

Another type of screening is the **enrichment** technique used when isolating auxotrophs. In one method the wild type strain is grown for many generations in minimal medium containing penicillin. Penicillin inhibits the growth of cell walls in dividing cells and so selectively inhibits the growth of the wild type (prototrophic) cells, at the same time increasing the proportion of non-dividing auxotrophs.

Mutations in other genes, such as those encoding the components of the replicational machinery, are more difficult to isolate and to study. These are **essential** genes, indispensable for the growth and survival of the cell, and the only mutations that can be detected are **conditional lethals**; these mutants can grow under one set of environmental conditions (the permissive conditions) but not under a different set (the restrictive conditions). Most frequently used are the **temperature-sensitive** (*ts*) mutants and these usually grow normally at 25°C but express the mutant phenotype at 42°C. Thus it is possible to propagate the mutant under the permissive conditions and to study the effect of the mutation under restrictive conditions.

6.8.2   Types of mutations

Mutations can be classed as either multi-site mutations or point mutations according to their effect on the nucleotide sequence of the genome. The **multi-site mutations** or macrolesions (known as chromosomal mutations in eucaryotes) involve major alterations in the nucleotide sequence and are of several different types (box 6.2). From a practical point of view the most important multi-site mutations are those caused by the insertion of an IS element into a gene, as these not only constitute a significant proportion of all spontaneous mutations (10–20 per cent of all Lac⁻ mutants, for example) but also the IS elements themselves are a frequent cause of the deletion and re-arrangement of adjacent nucleotide sequences (section 7.5). Multi-site mutations will not be considered further.

More frequent and more important are the **gene** or **point mutations**, which involve a single nucleotide pair (or sometimes a few adjacent nucleotide pairs) within a gene or a genetic control sequence. There are two types of point mutation, each of which

---

Box 6.2 The classification of multi-site mutations

(i) **Deletions**

A segment of DNA, varying from a few nucleotides to several kilobases, is missing and one or more genes are either missing or inactivated. Large deletions are usually lethal but small deletions are useful tools in genetic mapping (section 2.3.3).

(ii) **Rearrangements**

All the DNA is present but the nucleotide sequence has been rearranged; either a segment of DNA is inverted in relation to the rest of the DNA (**inversion**) or a segment of DNA has been excised from the chromosome and re-inserted at a new location (**translocation**). Rearrangements alter the linkage relationships between pairs of markers and can, for example, fuse the structural genes from one operon onto the promoter from a different operon.

(iii) **Duplications**

A particular segment of DNA is present twice. Although duplications are of frequent occurrence, they are inherently unstable as homologous recombination can occur between the two directly repeated segments, excising one and restoring the original sequence.

(iv) **Insertions**

A transposable element has been inserted into a gene, thereby inactivating it. These mutations can result from the transposition of an insertion sequence or transposon (chapter 7) or from lysogenisation by the mutator phage, Mu (chapter 8).

influences gene expression in a different way. These are: (i) **base substitution mutations** where a particular base pair has been replaced by a different base pair — these are referred to as **transitions**, when a purine replaces a purine and a pyrimidine replaces a pyrimidine, and **transversions**, where a purine replaces a pyrimidine and vice versa;

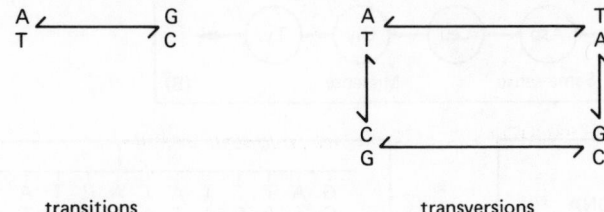

transitions   transversions

and (ii) **frameshift mutations** where one or two base pairs have either been added to or deleted from a coding sequence.

6.8.3 The effect of point mutations on protein synthesis

Point mutations can occur within either structural genes or genetic control sequences. A mutation in a structural gene affects the **function** of the gene product and usually alters the amino acid sequence of a polypeptide; however, some genes encode a species of messenger RNA that is not translated, and then mutation alters the nucleotide sequence in a molecule of transfer or ribosomal RNA. On the other hand, mutations in genetic control sequences affect the **rate** at which genes are transcribed and/or translated.

When a gene encodes a protein product four types of point mutation can be recognised (figure 6.9) and all exert their effect during translation:

(1)   **Samesense mutations** are usually detectable only by nucleic acid sequencing, for although the nucleic acid sequence of a codon is altered the same amino acid is still inserted into the polypeptide. This is because the genetic code is **degenerate** and most amino acids are encoded by more than one codon. For example, the codons GAU and GAC both specify aspartic acid so that an AT→GC substitution in the DNA that converts an mRNA codon from GAU to GAC (figure 6.9(B)) will not result in an amino acid substitution in the polypeptide. Any such mutation, which alters the nucleotide sequence without having a detectable effect on gene expression, is termed a **silent** mutation.

(2)   **Missense mutations** result when a base pair substitution in the DNA causes an amino acid substitution in the polypeptide. In figure 6.9(B) an AT→GC transition in the DNA has converted a UGG (tryptophan) codon to a GGG (glycine) codon; since these codons encode different amino acids, the result

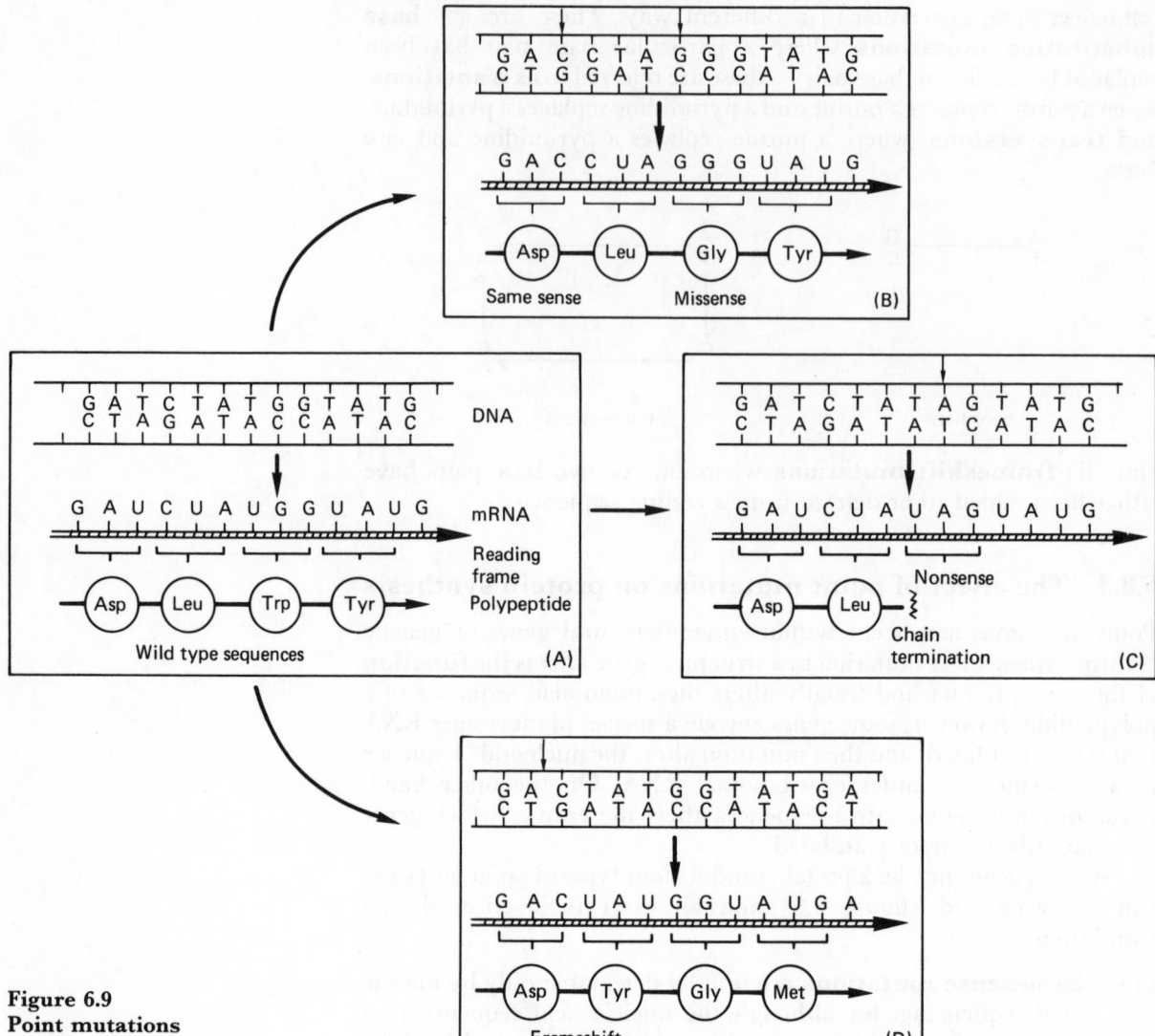

Figure 6.9
Point mutations

The figures show how a short
nucleotide sequence within a
structural gene is transcribed into
messenger RNA and translated into a
polypeptide. Note that the genetic
code is read in groups of three from a
fixed starting point so that the code is
always translated using an in phase
reading frame.
(A) The wild type nucleotide and
amino acid sequences.
(B) A samesense mutation has
occurred in the aspartic acid codon
and a missense mutation in the
tryptophan codon.
(C) A nonsense mutation has
occurred in the tryptophan codon.
(D) A frameshift mutation (− 1) has
occurred in the aspartic acid codon.

is a glycine for tryptophan substitution at the corresponding
position in the polypeptide.

The most important feature of a polypeptide is its ability to
fold up in a very specific way — in an enzyme, for example, this
creates the active sites that are essential for normal activity. Since
the amino acid sequence of a protein (its primary structure)
determines its final three-dimensional configuration (tertiary
structure), many amino acid substitutions result in a complete
or partial loss of activity and in expression of the mutant
phenotype. Other missense mutations are silent as, although they
cause amino acid substitutions, these do not critically affect the
conformation of the protein nor the functioning of the active
site(s).

(3)   **Nonsense mutations** are base pair substitutions that change
       a codon into a UAG, UAA or UGA chain termination triplet.
       When mutation generates a chain termination triplet (nonsense
       codon) in the middle of a gene (figure 6.9(C)), the translational
       machinery stops as soon as it reaches the nonsense codon so that
       only an inactive polypeptide fragment is produced; these codons
       are not translated because wild type cells do not contain a transfer
       RNA species capable of recognising them.

(4)   **Frameshift mutations** add or delete one or two base pairs
       from the message. This shifts the reading frame so that all codons
       from the one containing the extra or missing base to the end of
       the message are read out of phase; thus the polypeptide produced
       will contain the correct amino acid sequence up to the codon
       containing the frameshift, but thereafter every residue will be
       changed (figure 6.9(D)). Since a change in position of the
       reading frame frequently generates an in-phase nonsense codon
       between the site of the frameshift and the end of the gene, the
       polypeptides produced by many frameshift mutants are pre-
       maturely terminated before the end of the gene is reached.

6.9 Mutagenesis

Mutagenesis is the process of producing a mutation and when it occurs
naturally the mutations produced are known as **spontaneous** mutations
and they may be regarded as random changes in the nucleotide
sequence of an organism. The origin of spontaneous mutations is not
fully understood but many are undoubtedly due to occasional uncorrected
errors introduced during replication or repair (sections 6.2 and 6.6).
A feature of each of the four bases is that some of the hydrogen atoms
can change position and produce an alternative **tautomeric** form
which can then form an unconventional or mismatched base pair. For
example, the normal amino tautomer of adenine (figure 6.10) can only
pair with thymine, but when it is converted to the rare imino form it
pairs with cytosine. When this shift to the imino tautomer ($\overset{*}{A}$) occurs
in situ, then an $\overset{*}{A}$–C base pair will be produced at the next replication
(figure 6.11(A)); however, by the next replication the rare tautomer
will probably have reverted to the normal amino form and normal
base pairing will be resumed. The result is a G–C for A–T base pair
substitution in one of the two daughter molecules. In a similar way,
the misincorporation of the rare imino tautomer opposite a cytosine
(figure 6.11(B)) will lead to an A–T for G–C substitution.

Other spontaneous mutations are probably due to an increased
frequency of errors caused by the low levels of naturally occurring
background radiation.

The frequency of spontaneous mutation is very low, although every
gene has its own characteristic rate of mutation. In *E. coli*, a typical
gene has a forward-mutation rate between 2×10^{-6} and 2×10^{-8} per
gene per generation, while the back-mutation rate is usually considerably

Figure 6.10
The tautomers of thymine and adenine

Guanine, like thymine, shifts between the common keto and rare enol forms, while cytosine behaves like adenine and alternates between the common amino and rare imino tautomers.

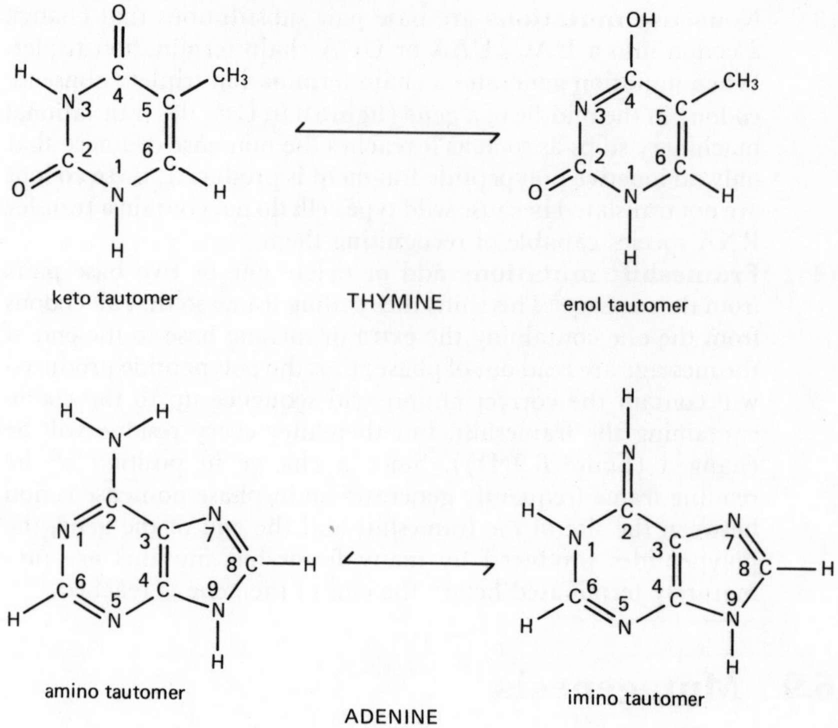

keto tautomer   THYMINE   enol tautomer

amino tautomer   ADENINE   imino tautomer

lower (2×10^{-7} to 2×10^{-9} per gene per generation). This is only to be expected since forward-mutation in an average gene can, in theory, involve any one of 1000 or so nucleotide pairs, while a true back-mutation can only occur by exactly reversing a previous forward-mutation and so involves a specific base pair change at one particular site within the gene.

However, the frequency of mutation can be considerably increased by exposing cells to a **mutagen** which produces **induced** mutations. Mutagens are commonly used in the laboratory to increase the likelihood of recovering mutants and we will consider how some of the different mutagens exert their mutagenic effect.

6.10   The molecular basis of mutagen activity

6.10.1   Base analogue mutagens

Base analogues are molecules which can substitute for the normal bases during DNA biosynthesis and, because they hydrogen-bond with a base on the template strand, they are not excised by the proofreading activity of DNA polymerases. The base analogues are comparatively weak mutagens and, in order to recover an adequate number of mutants, it is usual to grow bacteria in their presence for many generations.

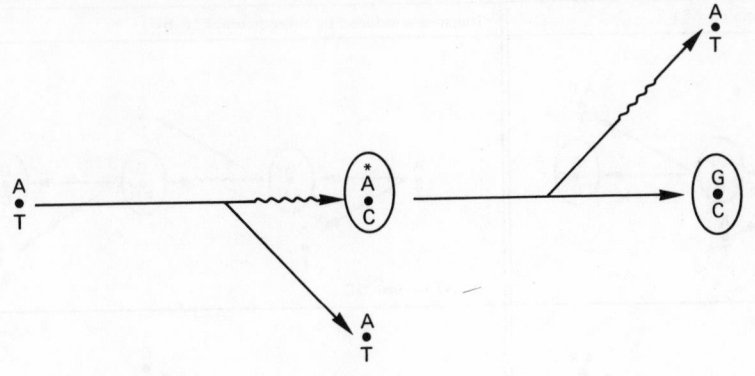

(A) AT → GC transition

Figure 6.11
**Transitions caused by
tautomeric shifts**

Occasionally, adenine shifts from the
common amino to the rare imino
form. An error of replication (above)
causes an AT → GC transition because
the rare imino tautomer pairs with
cytosine, while an error of
incorporation (below) is caused by
the mispairing of cytosine with the
free imino tautomer and results in a
GC → AT transition.

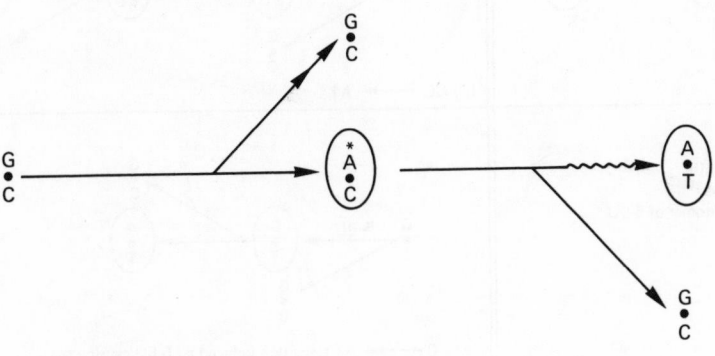

(B) GC → AT transition

〜〜〜〜   Tautomeric shift,

Å   imino tautomer of
         adenine

(i) 2-Aminopurine (2-AP)

This is an analogue of adenine and although it normally pairs with
thymine it can, very occasionally, pair with cytosine by the formation
of a single hydrogen bond. 2-AP is only rarely incorporated into DNA
in this way but when it is incorporated it acts as a potent mutagen.
2-AP can induce both the transitions AT → GC and GC → AT
(figure 6.12((A) and (B))), the former the result of an error of
replication (2-AP is incorporated opposite a T and subsequently pairs
with C) and the latter by an error of incorporation (2-AP is
misincorporated opposite a C).

(ii) 5-Bromouracil (5-BU)

5-Bromouracil, or 5-bromodeoxyuridine (5-BUdR), is an analogue of
thymine and the common tautomer pairs with adenine. Occasionally
this keto tautomer changes to the enol form which pairs with guanine;

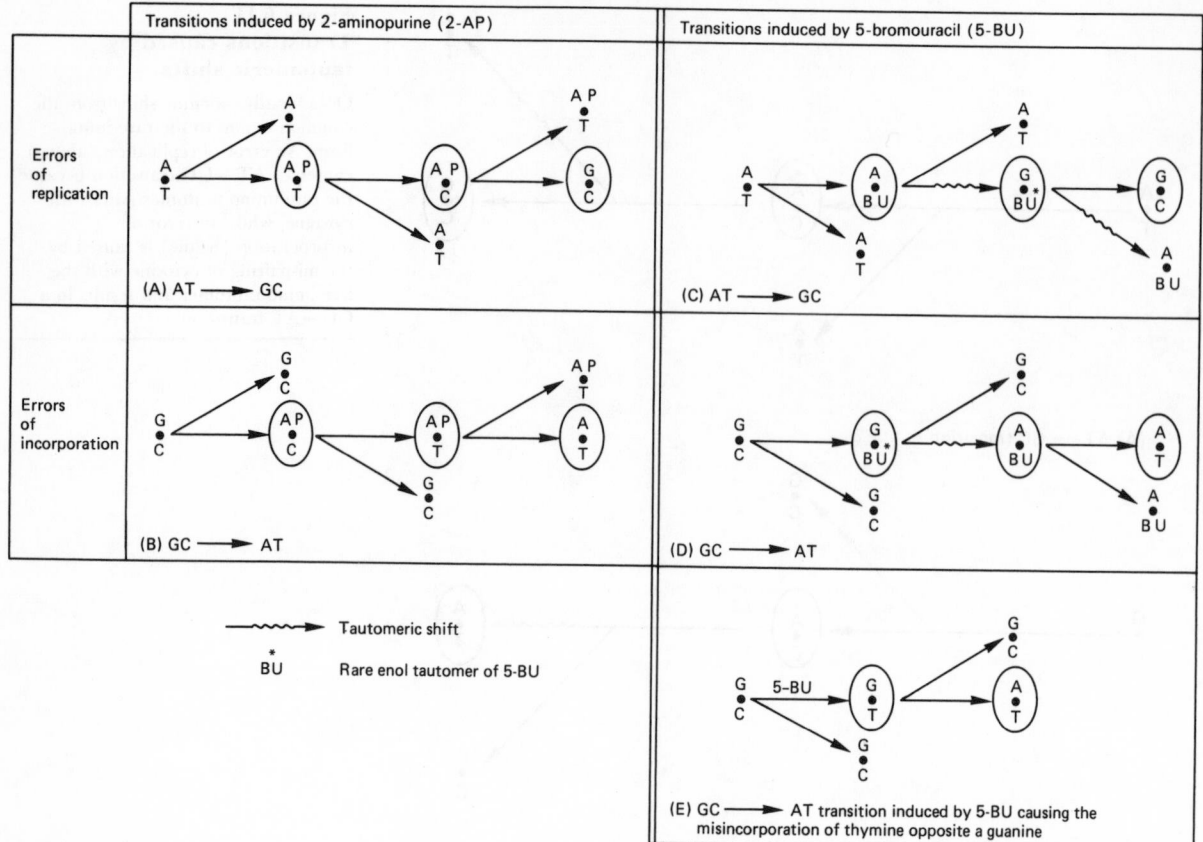

Figure 6.12
Transitions induced by
2-aminopurine and
5-bromouracil

These analogues induce both
AT → GC and GC → AT transitions;
2-AP because it occasionally mispairs
with cytosine ((A) and (B)) and 5-
BU because it occasionally changes to
the rare enol tautomer, which pairs
with guanine instead of adenine ((C)
and (D)). Note that 5-BU also
induces GC → AT transitions by
increasing the likelihood that thymine
will be misincorporated opposite a
guanine (E).

thus 5-BU, like 2-AP, causes both AT → GC and GC → AT transitions
(figure 6.12((C) and (D))). The only difference between 5-BU and
thymine is that the –CH$_3$ group on the C^6 of thymine is replaced by
bromine, but this small difference is sufficient to increase considerably
the frequency of tautomeric shifts from the keto to the enol form.

5-BU can also induce GC → AT transitions by another mechanism.
In most cells the concentrations of the four nucleoside triphosphates
are controlled by the level of thymidine triphosphate (TTP). One
feature of this complex regulatory system is that the synthesis of
deoxycytidine triphosphate (dCTP) is inhibited by excess TTP, or, if
it is present, by the nucleoside triphosphate of 5-BU. Thus when 5-BU
is present, TTP is synthesised at the normal rate but the synthesis of
dCTP is greatly reduced. This increases the ratio of TTP to dCTP
and makes more likely the mispairing of a thymine with a guanine; at
the next replication the misincorporated thymine will pair with an
adenine, producing a GC → AT transition in one of the daughter
molecules (figure 6.12(E)).

Although 2-AP and 5-BU induce both AT → GC and GC → AT
transitions, the former are preferentially induced by 2-AP and the
latter by 5-BU; however, in general, mutations induced by base
analogues can also be reverted by them.

6.10.2 Mutagens that modify the bases in DNA

Unlike the base analogues, which exert their mutagenic effect by replacing a normal base in **replicating** DNA, these chemicals alter the chemical structure of the nucleotide bases and so their action is independent of replication.

(i) Nitrous acid (HNO$_2$; NA)

This is a very potent mutagen which acts by oxidative de-amination. Its principal action is to convert guanine into xanthine, but since xanthine continues to pair with cytosine this change is not mutagenic. Less frequently, adenine is converted to hypoxanthine and cytosine into uracil, and since hypoxanthine pairs with cytosine and uracil pairs with adenine, both AT → GC and GC → AT transitions are induced (this is similar to the base analogue induced substitutions).

(ii) Hydroxylamine (NH$_2$OH; HA)

HA is an important mutagen because it reacts almost exclusively with cytosine to give a derivative ($\overset{*}{C}$) that pairs with A instead of G. As a consequence, HA only induces the transition GC → AT, and mutations induced by HA cannot be reverted by it (figure 6.13).

HA is both effective and specific when used to mutagenise transforming DNA or free phage, but its specificity is largely lost when it is used to treat intact bacterial cells; this is because it is a reducing compound and in an *in vivo* system it reacts with a wide variety of molecules producing compounds such as hydrogen peroxide, many of which are thought to be non-specific mutagens.

(iii) Alkylating agents

These are highly reactive compounds that transfer an alkyl group (CH$_3$–, CH$_3$CH$_2$–, etc.) to another molecule. When this transfer is to one of the nucleotide bases, the alkylated derivative may have a different pairing specificity. Thus O^6-alkylguanine can pair with thymine and O^4-alkylthymine can pair with guanine, leading to GC → AT and AT → GC transitions respectively.

A further effect of the alkylating agents on DNA is to cause strand breakage and depurination with the result that the error-prone SOS repair system is induced, and in turn this produces both transitions and transversions. This error-prone repair is an important source of mutations induced by the alkylating agents.

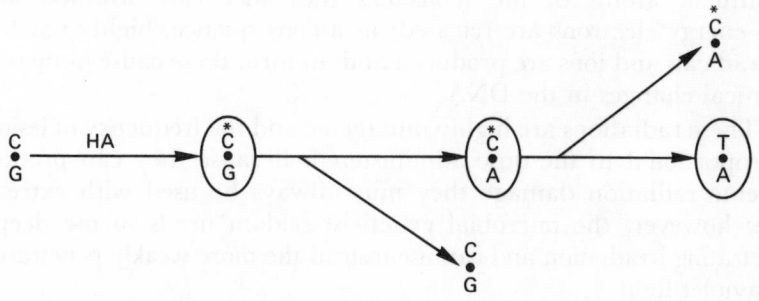

Figure 6.13
The induction of a GC → AT transition by hydroxylamine

Hydroxylamine modifies cytosine *in situ* by adding a hydroxyl group to its existing amino group. This altered base, hydroxylaminocytosine ($\overset{*}{C}$), can undergo a tautomeric shift enabling it to pair with adenine, so, ultimately, inducing a GC → AT transition.

Alkylating agents used as mutagens include the mustard gases, epoxides, ethylmethanesulphonate (EMS) and methylmethanesulphonate (MMS).

The alkylating agents are all very dangerous chemicals and must always be handled with the utmost care. Not only are they toxic but even at very low concentrations they are highly potent mutagens and many of them are also highly carcinogenic. Since they are extensively used in industry, they are a potential danger as pollutants of our environment.

6.10.3 Mutagens that bind to DNA

The acridines and acridine-like derivatives (proflavin, acridine orange, ICR191, etc.) are flat aromatic molecules, with the approximate dimensions of a normal base pair, which act by **intercalating** into double-stranded DNA by 'sliding' in between two adjacent base pairs along the double helix. This causes errors at the next replication and the most frequent result is that one (or sometimes two) base pairs are added to the DNA sequence; less frequently a base pair is deleted. The precise mechanism that produces these frameshift mutations is not known but these mutagens have a marked specificity for particular target sequences. Thus ICR191 and 9-aminoacridine recognise a repeating G sequence while 2-nitrosofluorene recognises both repeating G and repeating GC sequences.

These mutagens do not induce either transitions or transversions.

6.10.4 Electromagnetic irradiation

Electromagnetic radiations are widely used to induce mutations and they cause a wide variety of lesions in DNA. The genetically important types of radiation are ultraviolet light (approximate wavelength range from 100 to 10 nm), X-rays (10 nm to 1 nm), γ-rays (10^{-1} nm to 10^{-3} nm) and cosmic rays (below 10^{-3} nm), and their energies and penetrating powers are inversely proportional to their wavelengths.

The high-energy radiations (X-rays, γ-rays and cosmic rays) penetrate deeply into biological tissue and cause severe damage to the genetic material in two different ways. Firstly, single-strand and double-strand breaks are caused when quanta of energy hit the DNA; this is the direct effect. Secondly, and more importantly, there is an indirect effect. When these radiations pass through a cell, the constituent atoms of the molecules they meet are **ionised** and high-energy electrons are released; as a consequence, highly reactive free radicals and ions are produced and, in turn, these cause numerous chemical changes in the DNA.

These radiations are highly mutagenic and the frequency of lesions is proportional to the dose administered. Because they can produce extreme radiation damage they must always be used with extreme care; however, the microbial geneticist seldom needs to use deeply penetrating irradiation and can use instead the more weakly penetrating ultraviolet light.

Ultraviolet irradiation, although much less effective than ionising radiation, is still a fairly potent mutagen in microbial systems, and although it is only weakly penetrating it can easily pass through the cell membrane. The most effective wavelength is 254 nm, corresponding to the wavelength most strongly absorbed by the nucleotide bases.

Ultraviolet light, unlike high-energy radiation, is not ionising and it produces its effect by the excitation of orbital electrons, raising them to a higher state of energy; the primary effect of this is to induce the formation of pyrimidine dimers (section 6.2).

When bacterial cells are irradiated, the DNA is damaged and the SOS response is induced. This is important as most irradiation-induced mutations result from the replication errors introduced during the error-prone SOS repair process (section 6.7).

6.11 Hot-spots

Although we have described mutation as a random process, this is not quite true and some sites (that is, base pairs) within a gene are more susceptible to mutation than others; since there are only two possible base pairs (A–T and C–G), the frequency of mutation at these highly mutable sites, or **hot-spots**, must be a consequence of the adjacent nucleotide sequences. Hot-spots were first demonstrated by **Seymour Benzer** in 1961 in a remarkably extensive series of experiments with phage T4, in which he isolated over 2400 spontaneous mutants within the *rIIA* and *rIIB* genes. These *rII⁻* mutants mapped at 288 different sites, but each site did not mutate at the same frequency. At about one-half of the sites only a single mutation was observed and most of the other sites mutated between 2 and 20 times but there were two exceptional sites, one of which mutated 312 and the other 615 times. Hot-spots were also observed after treatment with chemical mutagens and it is of interest that these were in different positions from the hot-spots for spontaneous mutation. Furthermore, an additional 82 mutant sites were identified from among the induced mutants, so it is likely that at least some of these sites only mutate spontaneously at a very low rate, or perhaps not at all.

More recently, **J. H. Miller and his colleagues** have studied the distribution of spontaneous and induced UAG (amber) nonsense mutations within *lacI*. Mutations to amber can be generated at 37 different positions within *lacI* as the result of a single base pair substitution; two of these sites are hot-spots for GC→AT transitions and account for 23 per cent and 16.7 per cent of all the *lacI* amber mutations.

These hot-spots occur within the occasional symmetrical sequences

$$5' \quad \text{CĊAGG} \quad 3'$$
$$3' \quad \text{GGTĊC} \quad 5'$$

These sequences are recognised by DNA cytosine methylase (see box 6.1) and the two inner cytosines (*) are methylated to

5-methylcytosine. We have seen that cytosine can be spontaneously de-aminated to uracil (section 6.2) forming a G–U mismatch, but that this uracil is rapidly excised by the uracil-N-glycosylase system (section 6.5.2); in just the same way, 5-methylcytosine can be de-aminated to 5-methyluracil, better known as thymine, but since thymine is a naturally occurring base it cannot be excised by the N-glycosylase system and the G–T mismatch will persist to produce an A–T for G–C transition at the next replication.

Exercises

6.1 Both DP I and DP III have proofreading activity and can correct mismatched bases introduced during replication. Why, therefore, does the cell need so many further systems for repairing damage to its DNA?

6.2 What important features distinguish photoreactivation from all other repair systems?

6.3 Explain why irradiated cells slowly recover their ability to replicate when stored in the dark.

6.4 In cells lacking a functional excision repair system (*uvrABC*⁻), dimers cannot be excised; explain how cells cope with these unexcised dimers.

6.5 The *uvrABC* excision repair system of *E. coli* recognises damaged DNA by the distortions produced in the double helix. However, some forms of damage, such as certain mismatched base pairs, do not produce a distortion but are still repaired by a type of excision repair. How?

6.6 What is SOS repair and how does it differ from all other repair systems?

6.7 What are tautomers and how may a tautomeric shift result in a mutation?

6.8 Are mutations caused by acridine dyes more likely to be harmful to an organism than mutations induced by a base analogue? Give the reasons for your answer.

6.9 Compare and contrast the effects of base analogues and alkylating agents on the mutational process.

6.10 Distinguish transitions and transversions. Why are transitions the easier to explain in molecular terms?

6.11 The base analogues induce transitions in both directions. Would you expect

(i) HA to revert all mutations induced by 5-BU

(ii) HA to revert mutations induced by HA

(iii) 5-BU to revert mutations induced by HA?

6.12 The following sequence represents part of a structural gene and it is translated with the reading frame in the phase shown. What do you consider to be the likely effect on protein structure of frameshift mutations which (i) delete the 4th base pair from the left and (ii) insert an $\frac{A}{T}$ base pair between the 1st and 2nd base pairs?

```
• • • •A C G T A T C A G G• • • •
                                    DNA
• • • •T G C A T A G T C C• • • •


• • • •U G C A U A G U C C• • • •   mRNA
      └┬┘ └┬┘ └┬┘ └┬┘            reading frame
```

References and related reading

Cox, E. C., 'Bacterial mutator genes and the control of spontaneous mutation', *Ann. Rev. Genetics*, **10**, 135 (1976).

Drake, J. W., *Molecular Basis of Mutation*, Holden Day, San Francisco (1970).

Freese, E., 'Molecular mechanisms of mutation', in *Chemical Mutagens, Principles and Methods for their Detection*, (ed. A. Hollaender), Plenum, New York, p. 1 (1971).

Friedberg, E. C., *DNA Repair*, Freeman, New York (1984).

Haseltine, W. A., 'Ultraviolet light repair and mutagenesis revisited', *Cell*, **33**, 13 (1983).

Howard-Flanders, P., 'Inducible repair of DNA', *Scientific American*, (Feb.) **245**, 72 (1981).

Lindahl, T., 'DNA repair enzymes', *Ann. Rev. Biochem.*, **55**, 61 (1982).

Little, J. W. and Mount, D. W., 'The SOS regulatory system of *E. coli*', *Cell*, **29**, 11 (1982).

Miller, J., 'Mutational specificity in bacteria', *Ann. Rev. Genetics*, **17**, 215 (1983).

Roth, J. R., 'Frameshift mutations', *Ann. Rev. Genetics*, **8**, 319 (1974).

Walker, G. C., Marsh, L. and Dodson, L. A. 'Genetic analysis of DNA repair — inference and extrapolation', *Ann. Rev. Genetics*, **19**, 103 (1985).

Witkin, E., 'Ultraviolet mutagenesis and inducible DNA repair in *Escherichia coli*', *Bacteriol. Rev.*, **40**, 869 (1976).

7 TRANSPOSABLE GENETIC ELEMENTS

7.1 Introduction

Transposable genetic elements were first postulated by **Barbara McClintock** in the late 1940s to account for unusual patterns of inheritance of pigment distribution on the cobs of maize. During the next 20 years, McClintock, working at the Cold Spring Harbor Laboratories, New York, was able to show by painstaking genetic analysis that these **controlling elements** could not only move from one site in the genome to another but also could suppress gene activity, cause localised mutagenicity, induce chromosome breakage at the site where they were inserted, and regulate gene activity during development. However, the concept of mobile genetic elements was so contrary to the genetic wisdom of the 1950s that most geneticists rejected the concept and failed to realise the far-reaching consequences of McClintock's discoveries.

Although several studies in the early 1960s suggested that similar elements might be responsible for some genetic instabilities in bacteria and fungi, critical evidence was lacking, largely because the technology to enable the molecular isolation and characterisation of such elements was not then available. Nevertheless, it soon became clear that certain elements, like F and λ, could move in and out of the bacterial genome and so had some of the properties expected of transposable elements. However, true transposable elements, discovered in 1968, are unable to exist autonomously and can only replicate when they are a component part of an independent replicon.

True transposable elements in bacteria were discovered in 1968 by **Peter Starlinger and his group** at the University of Cologne and by **James A. Shapiro** at the University of Cambridge. They had isolated a number of mutants, mostly in the *gal* operon of *E. coli*, which had unusual properties. Not only did the effect of the mutation extend outside the mutant gene and reduce the expression of any downstream genes in the *gal* operon but, while some of the mutants only reverted to wild type at very low frequencies, others were very unstable and underwent further mutations at high frequency, rather like the controlling element mutations observed by Barbara McClintock. When the *gal*-operon DNA from these mutants was isolated and analysed it was found that the mutations were caused by the insertion into the operon of specific DNA sequences between 0.8 and 1.5 kb long, the length depending upon the particular element inserted. Furthermore, when the DNA from wild type reversions was similarly analysed, these additional DNA sequences were no longer present. These elements they called **insertion sequences** (IS).

At about this time, certain observations suggested that genes for antibiotic resistance were capable of transferring from an R plasmid to the bacterial chromosome and vice versa. The first direct evidence that this transfer took place and that it closely resembled the transposition of insertion sequences was only obtained in 1974, by **R. W. Hedges and A. E. Jacob**, working at the Royal Postgraduate Medical School in London. They found that whenever the gene for ampicillin resistance was transferred from one *E. coli* plasmid to another, the recipient plasmid always showed an increase in molecular weight of about 3×10^6 daltons (about 4.5 kb of duplex DNA); they suggested that the gene for ampicillin resistance was carried on a discrete genetic element that could transpose, and they called this element a **transposon**.

7.2   Insertion sequences

IS elements are characterised by their ability to transpose from one site in the genome to another and by their not encoding any function except those concerned with their own transposition. However, they do include transcriptional and translational start and stop signals, and so may have a dramatic effect on the expression of any more distal genes in the operon into which they are inserted. For example, the *gal* operon consists of three structural genes, *galE* (epimerase), *galT* (transferase) and *galK* (kinase), transcribed in that sequence from an operator–promoter region

$$PO-galE-galT-galK$$

and the insertion of an IS1 element into *galE* not only inactivates that gene but also prevents the expression of the downstream *galT* and *galK* genes.

This phenomenon is termed **polarity** and, because the IS1 element has such a severe effect on any downstream genes, the mutants they produce are referred to as **extreme polar mutants** (EPM). It was because of this extreme polar effect that IS elements were first detected.

With IS2, and probably with most other IS elements, these extreme polar effects occur because the element contains a rho-sensitive transcriptional termination sequence preceded by one or more in-phase nonsense codons; as a result the transcript initiated from the *gal* promoter will terminate and no *gal* sequences distal to the element will be transcribed. A further feature of IS2 is that it is only polar in one orientation, and in the alternative orientation transcription continues through the element and transcribes the distal *gal* genes. IS1, however, is polar in both orientations, and although it does not contain a rho-sensitive terminator sequence it probably exerts its influence in the same way as, like all IS elements, it contains a multiplicity of nonsense codons in all possible reading frames; thus when the ribosomes reach an in-phase nonsense codon they will dissociate from the messenger RNA and, in turn, this will promote the termination of transcription at the next rho-sensitive termination sequence within the distal *gal* genes.

With the exception of $\gamma\delta$, which is structurally and functionally related to Tn3, and IS101 which is only a vestigial element, all known IS elements have a similar structure and consist of a defined DNA sequence 0.7–1.5 kb long; their most conspicuous feature (figure 7.1) is that the two termini of each element are a pair of nearly perfect **inverted repeat sequences**. These inverted repeats are between 9 and 41 bp long (table 7.1) and are absolutely essential for transposition; they are probably the specific nucleotide sequences recognised by the element-encoded proteins, or **transposases**, which promote transpositional recombination.

All IS elements have a large **open reading frame**, a sequence of codons flanked by in-phase translational initiation and termination codons, which probably encodes the transposase. There is often a second and smaller open reading frame; this may be separate from the large reading frame, as in IS1 where it is known to encode a second protein required for transposition, or lie entirely within the larger reading frame but be transcribed in the opposite direction and from the complementary strand of DNA, as in IS2.

Figure 7.1
The structure of IS1

When IS1 transposes a 9-bp sequence of target site (TS), DNA is duplicated; these direct repeats are represented by the nucleotide sequences (i)–(i). Each terminus of IS1 is a 23-bp inverted repeat (IR) with 20 of the 23 nucleotide pairs corresponding exactly; these sequences are represented by (ii)–(ii).

IS1 has two open reading frames, *orfA* and *orfB*, which encode proteins involved in IS1 transposition. Note that the promoter for *orfA* lies within the left-hand inverted repeat and that *orfB* extends into the right-hand inverted repeat sequence. (Adapted from Machida, Y., Machida, C. and Craigie, R., *J. Molec. Biol.*, **177**, 229 (1984).)

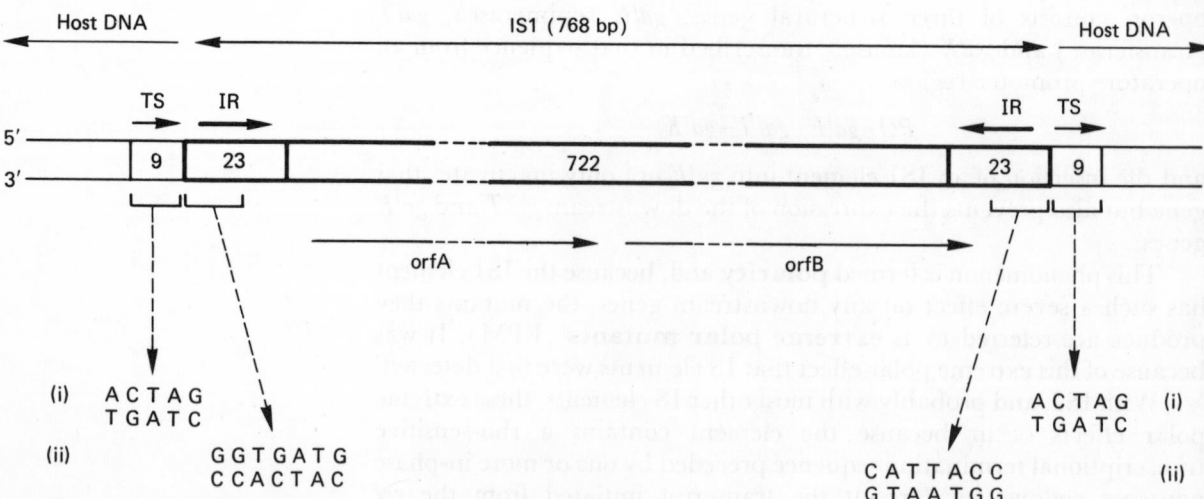

Table 7.1
Some E. coli insertion sequences

The inverted repeats are given in the form 20/23, meaning that 20 of the terminal 23 base pairs are identical.

Insertion sequence	Length (bp)	Number of copies on		Target site (bp)	Inverted repeats (bp)	Aberrations induced
		E. coli K12 chromosome	F			
IS1	768	4–10		9	20/23	co-integrates, deletions, inversions
IS2	1327	4–13	1	5	32/41	deletions
IS3	~1400	5 or 6	2	3–4	32/38	deletions
IS4	1426	1 or 2		11–13	16/18	deletions
IS5	1195	10 or 11		4	15/16	
IS30	1250	2–8		?.	23/26	deletions
γδ	~5700	0–3	1	5	35/35	co-integrates, deletions
IS10R	1057	0		9	18/18	co-integrates, deletions, inversions
IS50R	1534	0		9	8/9	co-integrates, deletions

Each transposed element is flanked by a short (3–13 bp) **direct repeat** of host DNA. This sequence identifies the target site on the host DNA and it is duplicated during the transposition process (section 7.4). For any given IS element, the length of the direct repeat is constant although the sequence within it is variable; this is because insertion occurs at sites that are selected more or less at random although, as with IS1, there may be a preference for regions of the genome with a high A–T content.

A feature of insertion sequences is that they can cause a variety of chromosomal aberrations, notably the deletion or inversion of DNA segments immediately adjacent to the inserted element and the formation of co-integrate molecules (table 7.1). The occurrence of these aberrations is largely dependent upon the transposition process and is considered in section 7.5.

Both IS2 and IS3 play an important role in the formation of Hfr strains of *E. coli*. The chromosome of *E. coli* carries multiple copies of IS2 and IS3, both of which are also present on the F plasmid and, by homologous recombination occurring between the copies of IS2 (or IS3) on the chromosome and on the F plasmid, an Hfr strain is formed with F inserted into the IS2 sequence on the bacterial chromosome (section 3.2.4).

7.3   Transposons

Transposons are distinguished from insertion sequences by their much larger size, usually between 2 and 20 kb, and by the presence of at least one gene which confers upon the host bacterium a heritable property. The most common and most intensively studied transposons carry genes conferring resistance to antibiotics but other characters may also be transposon-encoded; these include the genes for arginine biosynthesis (Tn2901), resistance to heavy metal ions (Tn4) and enterotoxin production (Tn1681).

7.3.1   Composite transposons

The first class of transposons comprises the **composite** transposons (table 7.2); these are relatively simple transposons, typically between 2.5 and 9.5 kb and having the generalised structure shown in figure 7.2(A). They consist of a central region usually encoding resistance to one or more antibiotics flanked either by a pair of identical IS elements or by a pair of sequences closely resembling an IS element; the latter are called IS-like modules. These flanking elements may be present as either inverted repeats (Tn5 and Tn10) or, more rarely, direct repeats (Tn6 and Tn9), but irrespective of this orientation the termini of each complete transposon will be a pair of short, inverted repeat sequences; this is because every IS or IS-like module has specific inverted repeat sequences at its inner and outer ends. At least one of the flanking IS or IS-like modules encodes the transposase protein(s) which acts upon the terminal inverted repeat sequences to promote

Table 7.2
Some E. coli transposons

Element	Length (kb)	Genetic markers		Flanking modules and orientation	Target site duplication (bp)
CLASS 1—Composite Transposons					
Tn5	5.7	kan^r	kanamycin resistance	IS50, inverted	9
Tn9	2.6	cam^r	chloramphenicol resistance	IS1, direct	9
Tn10	9.3	tet^r	tetracycline resistance	IS10, inverted	9
Tn1681	2.1	ent	enterotoxin (heat stable)	IS1, inverted	9
Tn2901	11.0	arg	arginine biosynthesis	IS1, direct	
CLASS II—Complex Transposons					
Tn1	5.0	amp^r	ampicillin resistance	38 bp inverted repeats	5
Tn3	4.9	amp^r	ampicillin resistance	38 bp inverted repeats	5
$\gamma\delta^*$	5.8	?		36/37 bp inverted repeats	5

*Not strictly a transposon since the only genes present are concerned with its own transposition, but included here because of its structural and functional homology with Tn3.

Figure 7.2
Class I and class II
transposons

A. Tn5 is a class I transposon. Each
element is flanked by a pair of IS50
elements in inverted orientations.
Although IS50L and IS50R differ at
only one nucleotide position, only
IS50R is fully functional; note the
short inverted repeats at the ends of
each IS50 module.

These IS-like modules flank a long
sequence of DNA containing a single
gene-encoding resistance to the
antibiotic kanamycin; the function of
the 2 kb of non-coding DNA is not
known.

B. Tn3 is a typical class II
transposon. The element is flanked by
38 bp inverted repeats and encodes
three genes. The *bla* gene encodes an
enzyme conferring resistance to
ampicillin; *tnpA* and *tnpR* encode a
transposase and a resolvase, both
essential for transposition. The *res*
site, located in the *tnpA–tnpR* intergenic
region, is where site-specific
recombination occurs during
transposition.

transposition; thus the complete transposon behaves and transposes in
just the same way as one of the flanking IS modules. These composite
transposons duplicate a 9 bp sequence when they insert at a new
position.

Soon after Hedges and Jacob had demonstrated the existence of
transposons, **Dennis Kopecko and Stanley Cohen** were able to show
that the two ends of a transposon consisted of nucleotide sequences
that are complementary to each other and in the reverse order. As a
consequence of the presence of these inverted repeats or **palindromes**,
a characteristic stem and loop structure forms when the DNA of a
plasmid carrying a transposon is denatured and allowed to re-anneal
(figure 7.3).

When composite transposons are flanked by a pair of IS1 elements
(such as Tn9) both IS1 modules are fully functional, can transpose as
independent units and either module can probably promote movement
of the transposon. However, in other composite transposons only the
right-hand module is fully functional and can actively promote
transposition (IS50R of Tn5 and IS10R of Tn10) while the second
module is either non-functional (IS50L) or has a reduced ability to
promote transposition (IS10L). Furthermore, it seems probable that
any pair of IS elements can cooperate to transpose any sequence lying

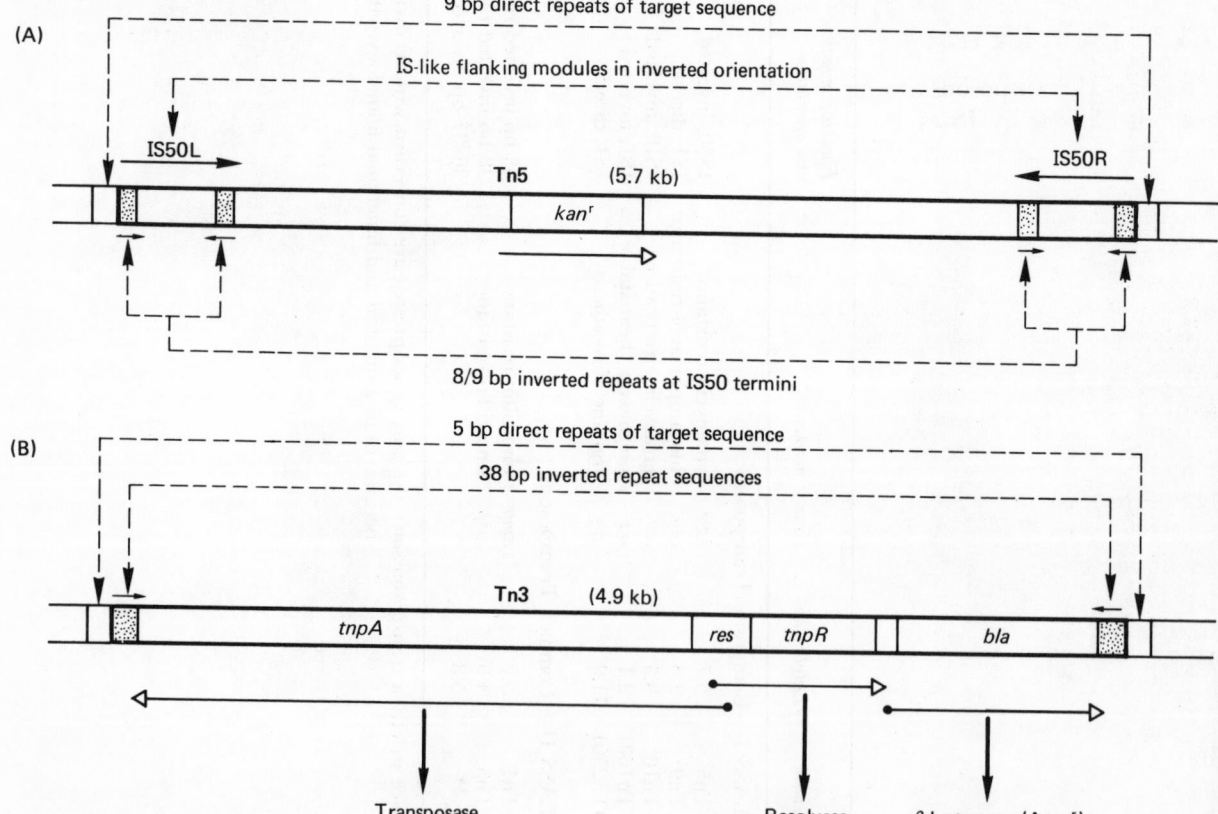

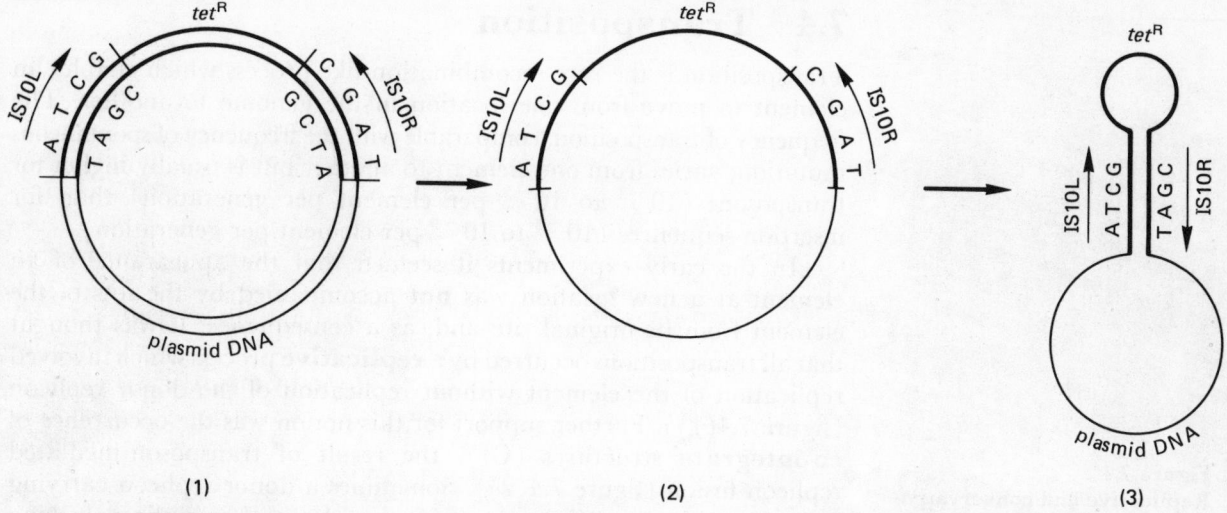

(1)   (2)   (3)

between them so that almost any gene can be made into a transposon by flanking it with two identical IS elements.

Thus the composite transposons behave in just the same way as IS elements but, because they usually carry genes for drug resistance, their presence and transposition is more easily monitored.

7.3.2 Complex transposons

The class II transposons are the complex or TnA type elements. These are not flanked by IS or by IS-like elements but by a pair of short, inverted repeat sequences, typically 35–40 bp long; these are equivalent to the ends of an IS element and are the specific sequences recognised by the transposase proteins. These IR sequences have no independent transpositional activity and, consequently, a TnA element can only transpose as a unit. Between these termini are one or more resistance genes and two genes encoding proteins essential for transposition. Like other transposable elements, these elements can transpose into many different sites on a molecule of recipient DNA, and whenever they do so they duplicate a 5 bp sequence of target DNA.

Tn3 (figure 7.2(B)), the most intensively studied class II element, has three genes encoding: (i) β-lactamase (*bla*), a penicillinase conferring resistance to ampicillin; (ii) a transposase protein (*tnpA*); and (iii) a resolvase protein (*tnpR*). Both the transposase and the resolvase are essential for transposition but the resolvase protein is of particular interest as it has two distinct functions. It acts by binding to three separate sites in the *res* (resolution) region (also known as the **IRS** or **internal resolution site**) located between the *tnpA* and *tnpR* genes where, on the one hand it promotes the site-specific recombination event that is the final step in the transposition process and, on the other hand, because this region also includes the promoter sequences for both *tnpA* and *tnpR*, it can act to repress the transcription of both these genes by preventing RNA polymerase from binding to the respective promoters.

Figure 7.3
Stem and loop structures

At the left (1) is a molecule of plasmid DNA carrying the tetracycline-resistance transposon Tn10 (heavy line); the inverted repeat IS10 sequences at each end of Tn10 are represented by arbitrary tetranucleotide sequences. If the plasmid DNA is denatured (2) and the resulting single strands allowed to re-anneal separately (3), stem and loop structures form and can be detected by electron microscopy. The heteroduplex segments arise because complementary base pairing occurs between the reverse and complementary sequences present on the same strand of DNA.

7.4   Transposition

Transposition is the rare recombination-like process which enables an element to move from one location in the genome to another. The frequency of transposition, comparable with the frequency of spontaneous mutation, varies from one element to another but is usually higher for transposons (10^{-3} to 10^{-5} per element per generation) than for insertion sequences (10^{-5} to 10^{-9} per element per generation).

In the early experiments it seemed that the appearance of an element at a new location was **not** accompanied by the loss of the element from its original site and, as a consequence, it was thought that all transpositions occurred by a **replicative** process which involved replication of the element without replication of the donor replicon (figure 7.4(1)). Further support for this notion was the occurrence of **co-integrate** structures (CI), the result of transposon-mediated replicon fusion (figure 7.4(2)). Sometimes a donor replicon carrying a transposable element becomes fused with another replicon lacking that element and, invariably, the resulting co-integrate carries two directly repeated copies of the element, one at each replicon–replicon junction. However, it is now clear that while some elements regularly transpose by a replicative process ($\gamma\delta$ and the TnA transposons), others normally transpose by a non-replicative or **conservative** process (figure 7.4(3)) where the element is cut from its location on the donor replicon and inserted into a target sequence on a different replicon (intermolecular transposition). It is probable that this results in destruction of the donor replicon and that the apparent retention of the element at its original site is a consequence of there being more

Figure 7.4
Replicative and conservative transposition

In replicative transposition the transposable element appears at a new location without being lost from its original site. It can give rise to either a simple insertion (1) or a co-integrate molecule (2). With conservative transposition the transposable element moves from its original site to a new site (3). Co-integrates cannot be formed by conservative transposition since the donor replicon is probably destroyed during the process.

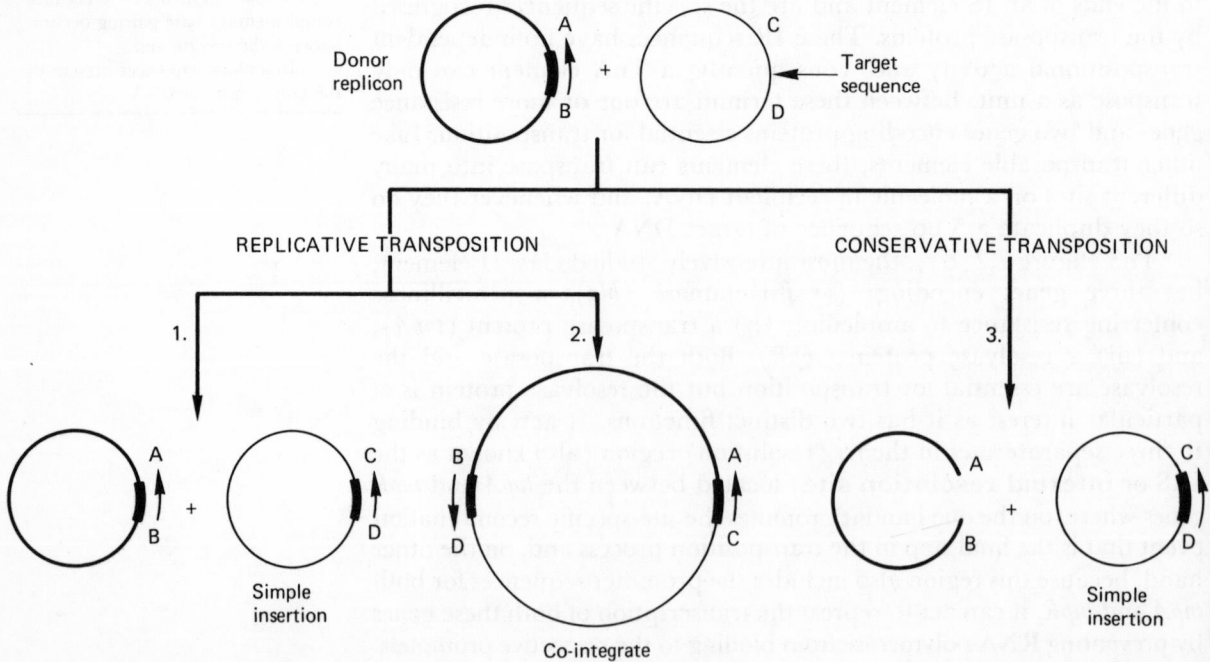

than one copy of the donor replicon in the cell at the time of transposition.

It is important to realise that transposition absolutely requires both the element-encoded transposase proteins and the specific nucleotide sequences at which these proteins act — within the terminal inverted repeats or at the internal resolution site. The *E. coli* system for general recombination (chapter 5) is not required, although general recombination can assist in the resolution of cointegrates; thus, as a general rule, transposition continues unimpeded in *recA*⁻ hosts.

7.4.1 The transposition of Tn3

Tn3, probably the best understood of the transposable elements, shares a common structure and mechanism of transposition with the other class II elements and with $\gamma\delta$. Intermolecular transposition, the most easily understood type of transposition, is a two-step replicative process (figure 7.5) and includes

(i) The formation of a co-integrate by replicon fusion. This only occurs when a functional *tnpA*⁺ gene product is present and the co-integrate always carries two copies of Tn3 in the same orientation; this clearly demonstrates the replicative nature of Tn3 transposition.

(ii) The co-integrate is resolved by the product of the *tnpR*⁺ gene into its two component replicons, each now carrying a copy of

**Figure 7.5
The two-step replicative transposition of Tn3**

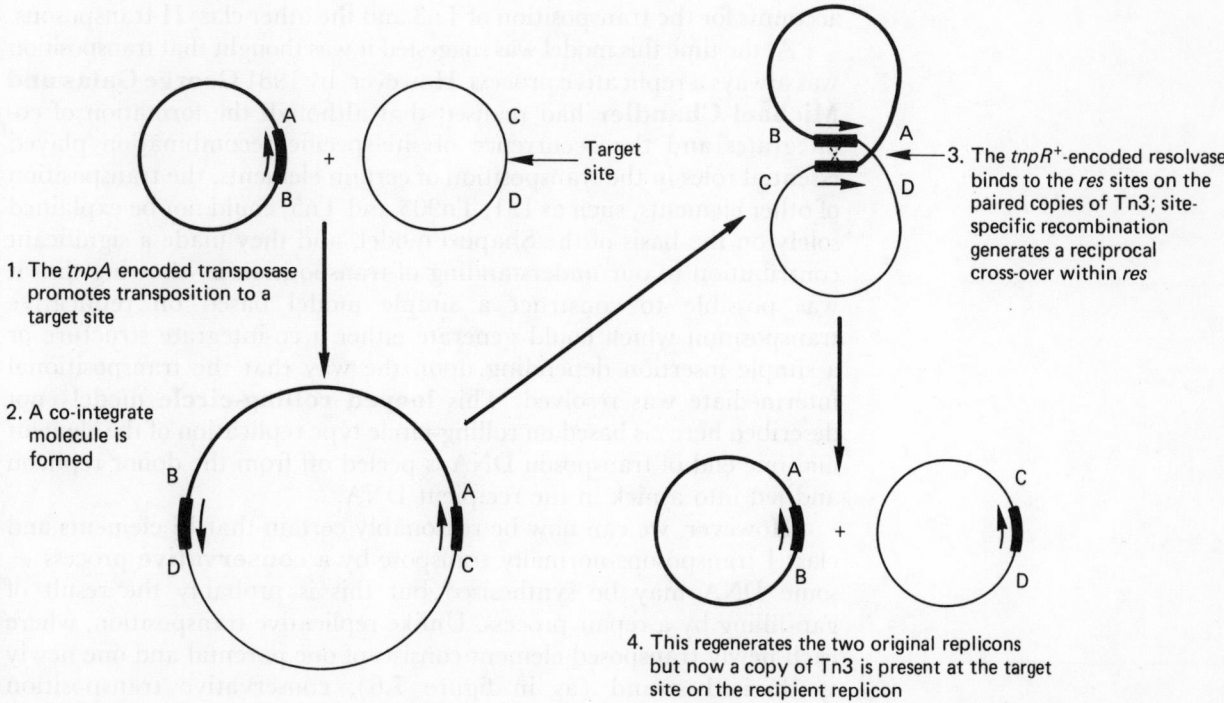

1. The *tnpA* encoded transposase promotes transposition to a target site

2. A co-integrate molecule is formed

3. The *tnpR*⁺-encoded resolvase binds to the *res* sites on the paired copies of Tn3; site-specific recombination generates a reciprocal cross-over within *res*

4. This regenerates the two original replicons but now a copy of Tn3 is present at the target site on the recipient replicon

Tn3. Resolution occurs by site-specific recombination, (sections 1.5.2 and 8.3), the $tnpR^+$-encoded resolvase recognising and binding to three sites within the 163 bp region between $tnpA$ and $tnpR$ where it promotes a single reciprocal cross-over.

These features of Tn3 transposition have been confirmed by the study of mutations occurring within $tnpA$, $tnpR$ and the res region. The $tnpA^-$ mutants are totally unable to transpose, while in $tnpR^-$ mutants the co-integrates form and accumulate since they cannot be efficiently resolved into the two component replicons. Note that the two directly repeated copies of Tn3 are regions of extensive homology and so, at least in principle, a co-integrate can be resolved by general recombination occurring anywhere along the length of the paired elements; however, this reaction is much less efficient than resolvase-promoted, site-specific recombination and it can only be detected in $recA^+$ hosts carrying a $tnpR^-$ transposon. Only when the host is $recA^-$ and Tn3 is $tnpR^-$ is co-integrate resolution completely abolished.

7.4.2 The mechanism of transposition

Several very different types of model have been suggested to explain transposition in molecular terms but at present there is little real evidence to indicate the true molecular basis of the process; one of the more plausible and widely accepted schemes, based largely on a model suggested by **James Shapiro** in 1979, is shown in figure 7.6. According to this scheme there is replication of the element without replication of the donor molecule, and co-integrates are **obligatory** intermediates in the formation of simple insertions. This model very satisfactorily accounts for the transposition of Tn3 and the other class II transposons.

At the time this model was suggested it was thought that transposition was always a replicative process. However, by 1981 **George Galas and Michael Chandler** had realised that although the formation of co-integrates and the occurrence of site-specific recombination played essential roles in the transposition of certain elements, the transposition of other elements, such as IS1, Tn903 and Tn5, could not be explained solely on the basis of the Shapiro model, and they made a significant contribution to our understanding of transposition by showing that it was possible to construct a simple model based on replicative transposition which could generate either a co-integrate structure or a simple insertion depending upon the way that the transpositional intermediate was resolved. This **looped rolling-circle** model (not described here) is based on rolling-circle type replication of the element and one end of transposon DNA is peeled off from the donor replicon and fed into a nick in the recipient DNA.

However, we can now be reasonably certain that IS elements and class I transposons normally transpose by a **conservative** process — some DNA may be synthesised but this is probably the result of gap-filling by a repair process. Unlike replicative transposition, where each newly transposed element consists of one parental and one newly synthesised strand (as in figure 7.6), conservative transposition

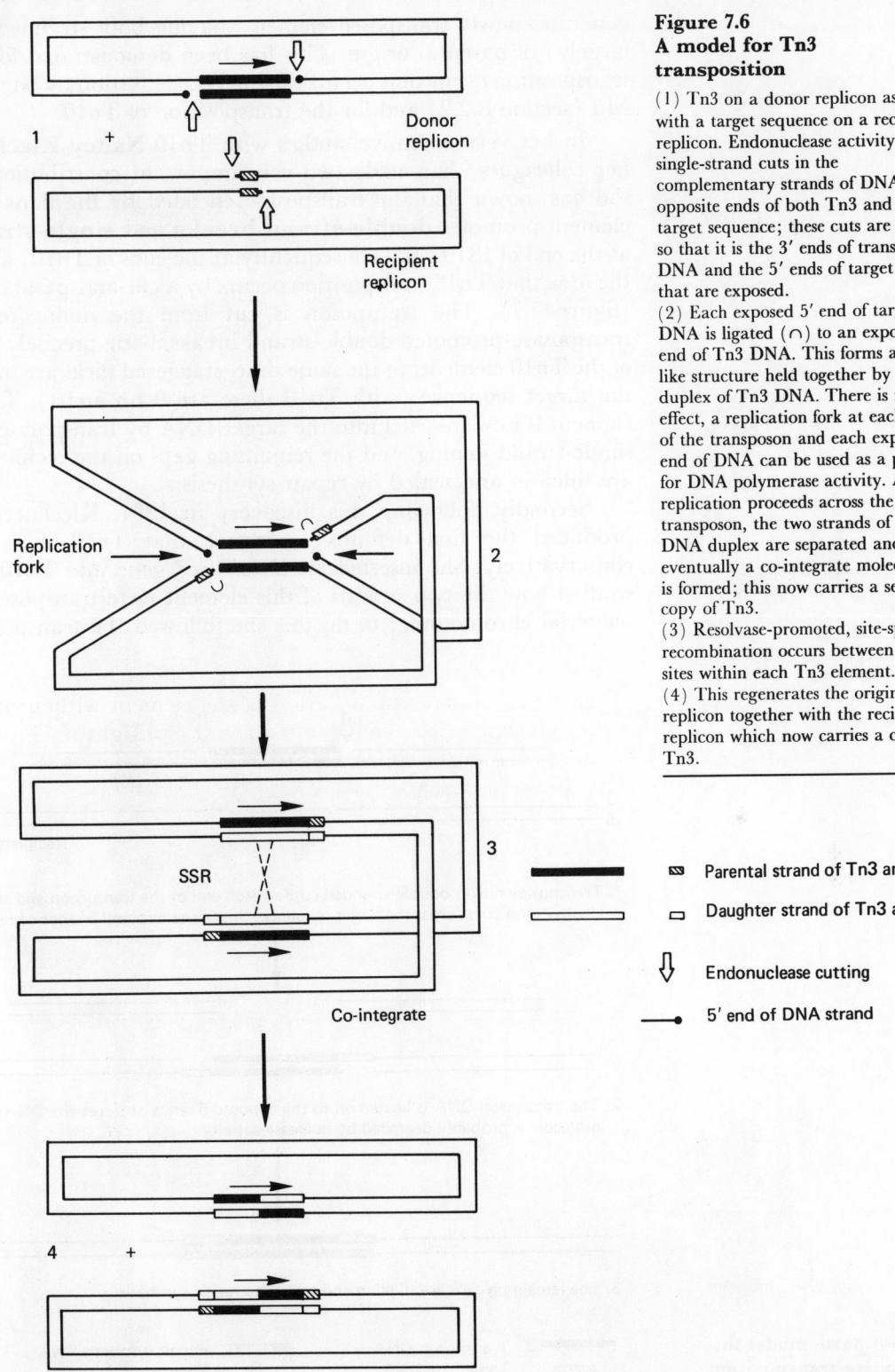

1 +

Donor
replicon

Recipient
replicon

Replication
fork

2

SSR

3

Co-integrate

4 +

Simple insertion

Figure 7.6
A model for Tn3 transposition

(1) Tn3 on a donor replicon associates with a target sequence on a recipient replicon. Endonuclease activity makes single-strand cuts in the complementary strands of DNA at opposite ends of both Tn3 and the target sequence; these cuts are made so that it is the 3′ ends of transposon DNA and the 5′ ends of target DNA that are exposed.

(2) Each exposed 5′ end of target DNA is ligated (∩) to an exposed 3′ end of Tn3 DNA. This forms a cross-like structure held together by the duplex of Tn3 DNA. There is now, in effect, a replication fork at each end of the transposon and each exposed 3′ end of DNA can be used as a primer for DNA polymerase activity. As replication proceeds across the transposon, the two strands of the DNA duplex are separated and eventually a co-integrate molecule (3) is formed; this now carries a second copy of Tn3.

(3) Resolvase-promoted, site-specific recombination occurs between the *res* sites within each Tn3 element.

(4) This regenerates the original donor replicon together with the recipient replicon which now carries a copy of Tn3.

▨ Parental strand of Tn3 and target DNA

▢ Daughter strand of Tn3 and target DNA

⇩ Endonuclease cutting

•— 5′ end of DNA strand

generates newly transposed elements having both strands wholly (or largely) of parental origin. This has been demonstrated for both the transposition event that occurs during lysogenisation by bacteriophage Mu (section 8.2.2) and for the transposition of Tn10.

In her very extensive studies with Tn10 **Nancy Kleckner** (and her colleagues) has made two very important contributions. Firstly, she has shown that the transposase encoded by the flanking IS10R element promotes **double-strand breaks** and **single-strand joins** at the end of IS10 (and consequently at the ends of Tn10) and favours the idea that Tn10 transposition occurs by a cut-and-paste mechanism (figure 7.7). The transposon is cut from the donor replicon by transposase-promoted double-strand breaks made precisely at the end of the Tn10 element; at the same time, staggered nicks are made across the target sequence (with Tn10 these are 9 bp apart). The excised element is now inserted into the target DNA by transposase-mediated single-strand joining, and the remaining gaps on the recipient strands are filled in and sealed by repair synthesis.

Secondly, following this discovery in 1984, Kleckner, in 1986, produced the first definitive evidence that Tn10 does transpose conservatively. She inserted an *E. coli lacZ* gene into Tn10 and then studied how the two strands of this element were transposed into the bacterial chromosome; to do this she followed the transposition of a

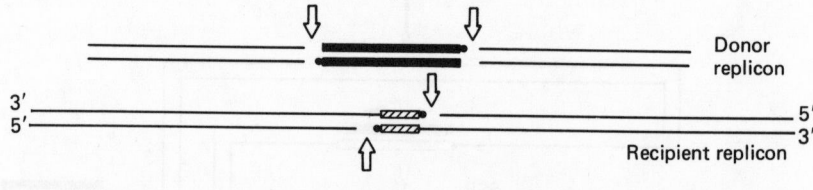

1. Transposase makes double-stranded cuts at each end of the transposon and staggered single-strand cuts across the target sequence; this leaves exposed 5' ends of target site DNA

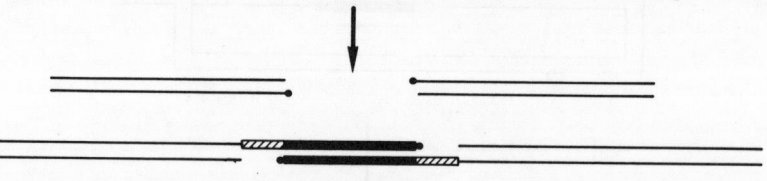

2. The transposon DNA is ligated on to the exposed 5' ends of target site DNA; the donor molecule is probably degraded by nuclease activity

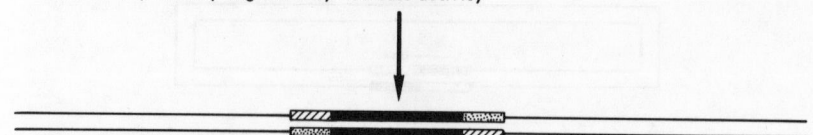

3. The remaining gaps are filled in and sealed by repair synthesis

Figure 7.7
A cut-and-paste model for
conservative transposition

heteroduplex Tn10 element, where one strand included the wild type *lacZ*⁺ sequence and the other a *lacZ*⁻ base substitution mutation. She argued that conservative transposition would yield transposition products containing **both** strands of genetic information so that *lacZ*⁺ and *lacZ*⁻ would segregate at the next replication, whereas semi-conservative replicative transposition would yield products containing information from only one strand (*lacZ*⁺) or the other (*lacZ*⁻). The pattern of marker recovery was exactly as expected if both strands of Tn10 were cut out from the donor molecule and integrated directly into the bacterial chromosome. The protocol for these experiments is set out in box 7.1.

A feature of the transposition process is the duplication of the target site sequence at each end of the transposed element. All the models for transposition explain this by assuming that staggered single-strand nicks are made across the target site and that the element is, in effect, inserted between the two single-stranded ends of the target sequence (as shown in figures 7.6 and 7.7).

However, it is clear that some IS elements and some class I transposons do form co-integrates (IS1 and IS903, for example) although their frequency is much lower (1–5 per cent) than the frequency of simple insertions; this suggests that these elements can sometimes undergo replicative transposition. Unlike Tn3, these elements do not have a site-specific recombination system and so the co-integrates must, if they are to be resolved, be resolved by general recombination; it is highly significant that these co-integrates do **not** accumulate in *recA*⁻ host cells as would be expected if all or most of the simple insertions were produced by the resolution of co-integrates. This raises the possibility of there being two transpositional pathways, and both **Eiichi Ohtsubo** and **Kiyoshi Mizuuchi** have suggested that co-integrates and simple insertions might be formed directly from a common transposition intermediate;

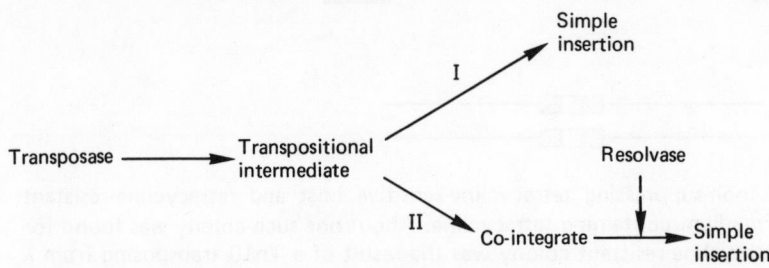

Thus the class II transposons would always proceed along pathway II, while the class I transposons and the insertion sequences would normally follow pathway I.

In 1985 Mizuuchi was able to show that simple insertions, like co-integrates, could arise directly from the transpositional intermediate predicted by the Shapiro model if, firstly, the replication proteins fail to assemble and form replication forks at each end of the element (figure 7.8) and, secondly, nuclease activity cuts and degrades the

Box 7.1 The method used by Nancy Kleckner to show that Tn10 transposes by a conservative mechanism

1.

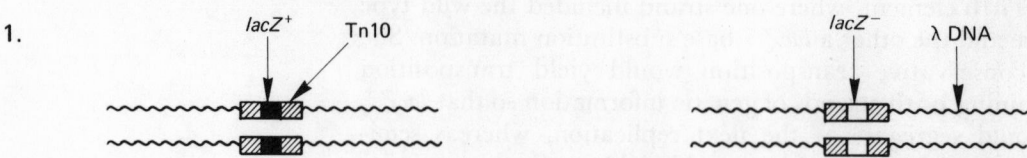

Recombinant DNA technology was used to construct two λ phages carrying Tn10 elements which, in turn, had *lacZ*⁺ or *lacZ*⁻ inserted into them. These phages were isogenic except for a single base pair mismatch within *lacZ*.

2.

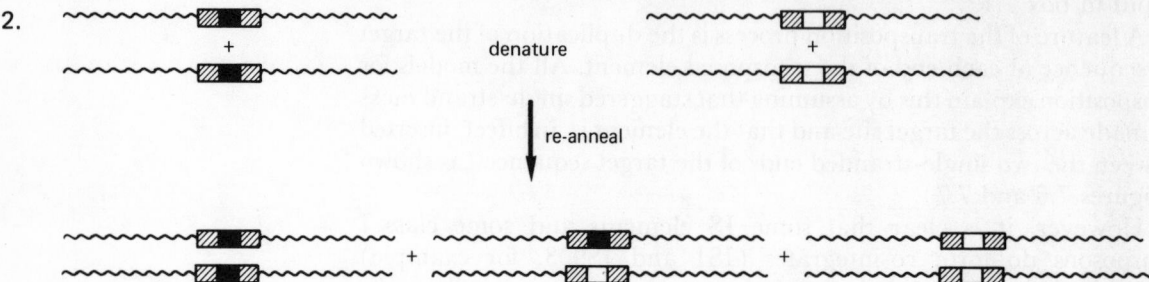

The two types of λ DNA were denatured and the single strands mixed and allowed to re-anneal, forming heteroduplex and both types of homoduplex molecules.

3. The molecules prepared by re-annealing were packaged *in vitro* into heads, forming infective λ particles. Note that the λ genomes used in these experiments were integration (*int*⁻) and repression (*c*⁻) deficient and had an amber mutation in the *P* (replication) gene; thus, when used to infect a non-suppressing host, they could neither integrate nor replicate.

4.

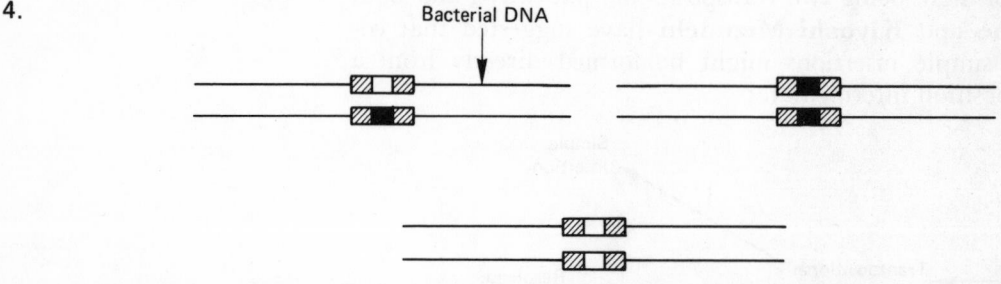

5. The phages were used to infect a non-suppressing tetracycline-sensitive host and tetracycline-resistant colonies selected on Xgal indicator medium containing tetracycline. About one such colony was found for every 10⁵ phage genomes. Each tetracycline-resistant colony was the result of a Tn10 transposing from λ to the bacterial chromosome.

6.

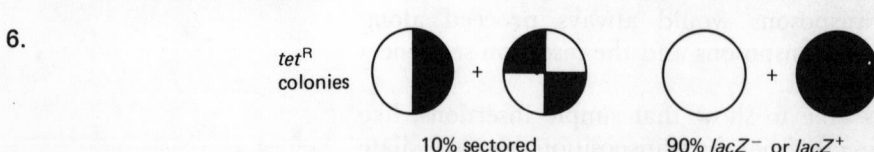

On Xgal medium, *lacZ*⁺ colonies are blue and *lacZ*⁻ colonies are white. About 10 per cent of the tetra-cycline-resistant colonies were sectored and so had segregated *lacZ*⁺ and *lacZ*⁻. The simplest explanation is that both strands of the heteroduplex Tn10 elements were transposed into the bacterial chromosome.

These experiments are fully described in Bender, J. and Kleckner, N., *Cell*, **45**, 801 (1986).

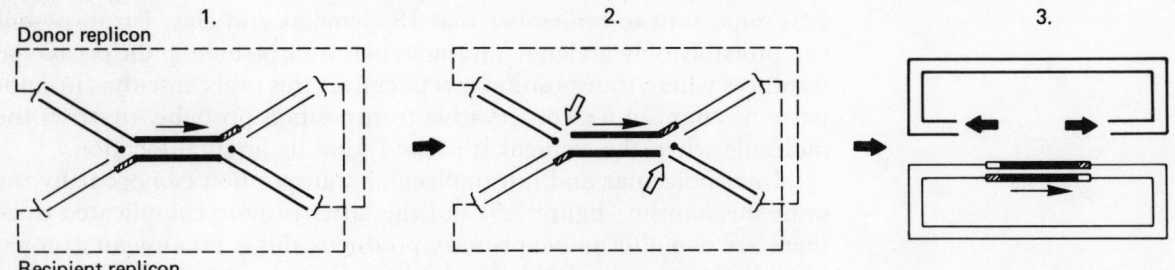

1.

2.

3.

Donor replicon

Recipient replicon

The transpositional intermediate predicted by the Shapiro model (figure 7.6)

The replication proteins have failed to assemble and to form replication forks at each end of the element. Endonuclease cuts the strands of donor DNA exactly at the ends of each element

After the resulting single-strand gaps have been filled in by repair synthesis, the recipient replicon will contain a copy of the element flanked by the directly repeated target site sequence. The linear molecule of donor replicon DNA will be unable to replicate and be degraded by exonuclease activity

⇨ Endonuclease   ➡ Exonuclease

strands of donor DNA at each end of the transposon. This produces a recipient replicon carrying the original transposon and a linear molecule of donor replicon DNA which would be rapidly degraded by nucleases.

There is another explanation for the occasional co-integrates found with certain IS elements and class I transposons. Just as a co-integrate can be resolved into its component replicons by a cross-over between the two directly repeated copies of the element (as in figure 7.5), so the reverse is also possible and crossing-over could fuse together two different replicons provided that each carried a copy of the same element. This would be dependent on the $recA^+$ system of general recombination since IS elements and class I transposons lack a site-specific recombination system. However, since these rare co-integrates continue to form in $recA^-$ hosts, it would seem more likely that insertion sequences and class I transposons do occasionally undergo replicative transposition.

Figure 7.8
An alternative model for producing a simple insertion

Note that endonuclease could also cut the strands of recipient DNA at the target sequence—element junctions; this event would remain undetected as it would reconstitute the donor replicon and result in the destruction of the recipient replicon.

7.5 Chromosomal aberrations induced by transposable elements

Another characteristic feature of many transposable elements is that the types of event involved in transposition can also generate a limited variety of genetic rearrangements (table 7.1).

The simplest rearrangement is the co-integrate formed when transposition occurs from one replicon to another — in other words during intermolecular replicative transposition (figure 7.5 and section 7.4.2). Other rearrangements, such as adjacent deletions and inversions, are possible following intramolecular transposition when an element transposes from one location to another on the same replicon.

It is important to remember that IS elements and class I transposons can probably only undergo intramolecular transposition in the occasional instances where transposition is replicative; this is because the cut-and-paste mechanism for conservative transposition probably destroys the molecule when the element is excised from its original location.

Intermolecular and intramolecular transposition can occur by the same mechanism (figure 7.6) but the latter is more complicated since there are two alternative primary products; this is because an exposed 5′ end of target site DNA can be ligated on to either of the exposed 3′ ends of transposon DNA. With one pattern of ligation the primary product is a complete molecule with two copies of the element in **opposite** orientations (figure 7.9(a)), and with all the genetic markers between one end of the transposable element and the target site in the opposite or **inverted** order; only if site-specific or homologous recombination occurs between the two copies of the element can the correct gene order be restored. This inversion type product is only rarely observed following the transposition of IS elements, but is more frequently found after the transposition of transposons such as Tn10. The alternative pattern of ligation results in two incomplete circular molecules each with a copy of the element (figure 7.9(b)); only the molecule with the original replication origin (*ori*) can replicate and be transmitted to the daughter cells so that, in effect, all the genes between one end of the element and the target site have been lost or **deleted**; this is known as **adjacent deletion**. Note that the element itself is retained intact and can continue to promote further deletions. When a transposable element is present, the frequency of deletions in the vicinity of the element is increased by 2–3 orders of magnitude.

Two other types of rearrangement involve transposable elements

Figure 7.9
Transposon-induced
chromosomal aberrations

Intramolecular transposition can produce either (a) a complete circular molecule with a second copy of the transposon but with all the markers between one end of the transposon and the target site in an inverted orientation, or (b) two incomplete circular molecules; only one molecule has a replication origin and can replicate, and this has the sequence ABC deleted.

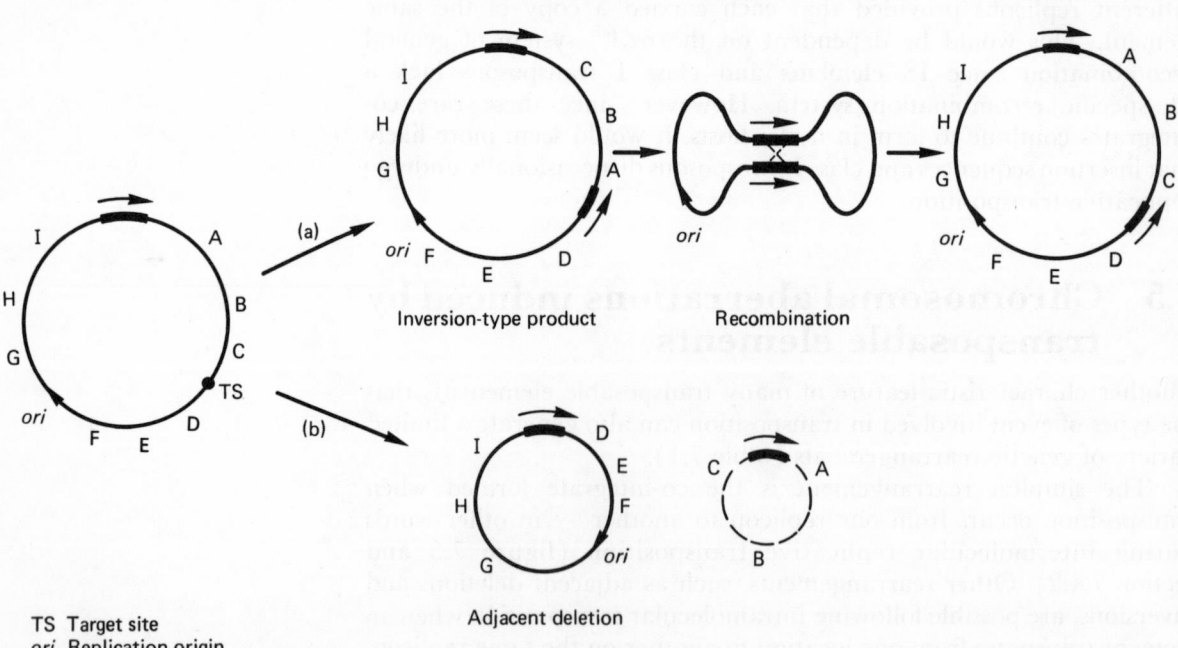

TS Target site
ori Replication origin

Inversion-type product   Recombination

Adjacent deletion

but are **not** promoted by transposition. Precise excision results in the loss of the transposable element together with one of the target site sequences, so reconstructing the original nucleotide sequence. This is a rare event (10^{-7} to 10^{-9} per bacterium per generation) occurring at a much lower frequency than transposition, and most simply could involve a cross-over between the duplicated target sequences flanking the element. The remaining type of rearrangement is imprecise excision which deletes part of the element and, sometimes, some of the adjacent host sequences as well; it is thought to be the result either of rare illegitimate crossing-over or of crossing-over between two directly repeated sequences, at least one of which lies within the element.

7.6   Factors affecting transposition

The frequency of transposition and of other transposition-related events is influenced by both genetic and environmental factors. We have already seen an example of genetic influence where, in Tn3, mutations which result in a loss of transposase (*tnpA*) activity abolish the formation of co-integrates while resolvase (*tnpR*) mutants are unable to resolve co-integrates into their component replicons. In a similar way, mutations within the inverted repeat termini of insertion sequences can affect the binding of transposase and so decrease the frequency of transposition. The influence of environment can be illustrated by the effect of temperature on the frequency of deletions promoted by IS1 and IS2; in this instance the frequency of deletions can be increased by up to 3 orders of magnitude by decreasing the temperature from 42°C to 32°C. However, for any given set of genetic and environmental conditions, the frequency of transposition and of other transposition-related events remains more or less constant, **suggesting that the activity of each element is under the control of a regulatory mechanism**. Until just recently, very little was known about this regulation but it is now clear that the activity of many transposable elements can be modulated by DNA methylation (see box 6.1).

7.6.1   The regulation of transpositional activity by DNA methylation

In 1985 **Nancy Kleckner and her coworkers** discovered that in *dam⁻* mutants of *E. coli* there is a 10-fold to 20-fold increase in the frequency of Tn10 transposition and Tn10-promoted deletion formation; these mutants lack DNA adenine methylase and are unable to methylate the ademine bases in the palindromic sequence 5′ GATC 3′
3′ CTAG 5′. IS10R contains only two GATC sequences, one of these lying within the promoter for the transposase gene (figure 7.10) and the other within the transposase binding site at the inner end of the element, and Kleckner has shown that the increased frequency of transposition is due to a loss of methylation at these two sites. This lack

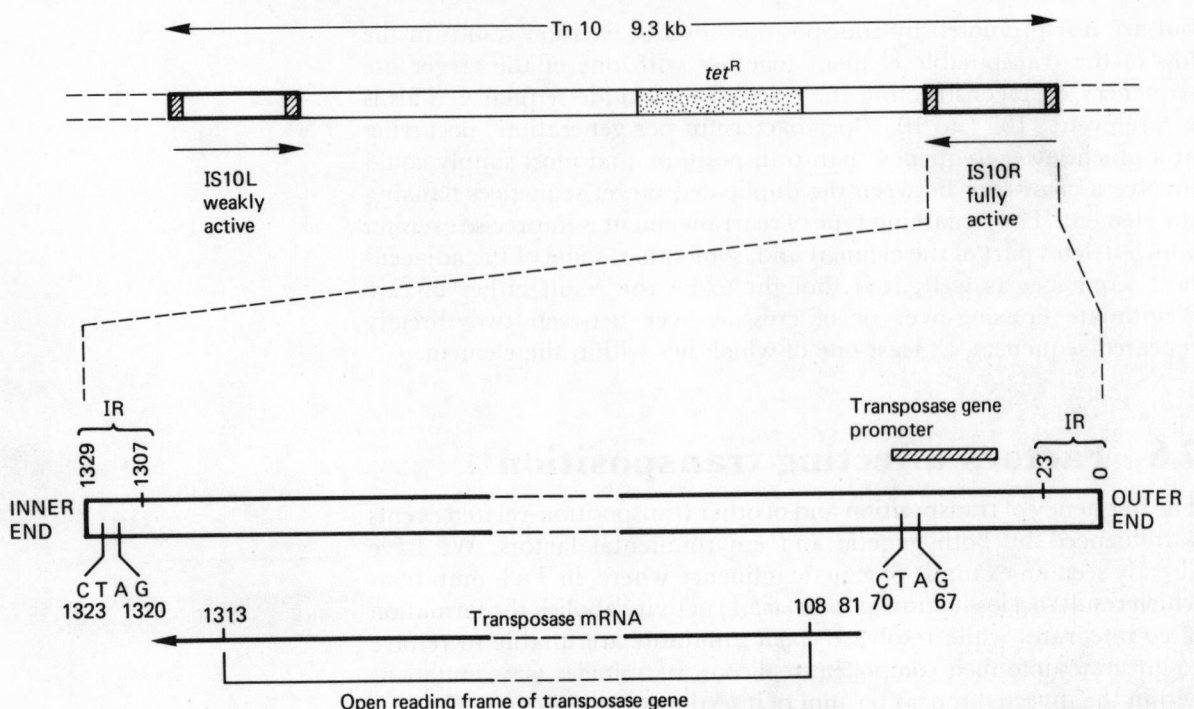

Figure 7.10
The structures of Tn10 and of the flanking IS10R element

IS10L and IS10R differ by at least 16 base pairs and only IS10R is fully functional. IS10R contains two *dam* methylation sites, one at 67–70 within the Pribnow Box of the transposase promoter and the other at 1320–1323 within the inner transposase binding site. The nucleotides of IS10 are numbered from the outer ends of each element. (Adapted from Roberts, D., Hoopes, B. C., McClure, W. R. and Kleckner, N., *Cell*, **43**, 117 (1985).

of methylation has a two-fold effect; firstly, demethylation of the promoter allows transcription of the transposase gene and the production of transposase and, secondly, demethylation of the transposase binding site allows the transposase to bind more effectively at the inner end of IS10. Kleckner also found that the frequency of IS10R transposition was even more strongly activated when the DNA was hemimethylated (that is, methylated in one strand only). This is the situation found *in vivo* just after the replication fork has passed over and replicated a Tn10 element, when each copy of the element will be transiently hemimethylated on opposite strands. It seems likely that the transposase is only synthesised and that transposition can only occur during this brief period immediately following replication and preceding methylation of the newly synthesised daughter strands.

Note that the three events monitored by Kleckner are all transposition-related, but that they involve different pairs of recognition sites within the symmetrical terminal inverted repeat sequences of IS10L and IS10R; transposition of the complete element (Tn10) requires transposase activity at the **outside** ends of both IS10L and IS10R, the transposition of IS10R requires this activity at the **outside** and **inside** ends of the same IS10 element, while deletion formation requires activity at the **inside** end of one IS10 element and at the **outside** end of the other element. These differences satisfactorily explain why these three events occur at different frequencies and how it is that events involving an inside end of IS10R (IS10R transposition and Tn10-promoted deletion formation) are more strongly stimulated by *dam⁻* than events involving only the two outer ends (Tn10 transposition).

The *dam*⁻ mutations also increase the frequencies of transposition of Tn5 and Tn903; this also appears to be due to increased promoter activity resulting from a lack of methylation at a GATC site within the promoters for the transposase genes in the flanking IS50 and IS903 elements.

7.7 The uses of transposable elements in molecular biology

Transposons and insertion sequences, because of their unusual properties, are very valuable tools for the use of geneticists and molecular biologists alike, and in this section we will summarise some of their special uses (see also section 8.2.5 on the uses of bacteriophage Mu). Although, as we have seen, insertion sequences and transposons have many properties in common, it is frequently more convenient to use the latter; this is because the resistance genes carried by transposons not only provide selectable markers but also enable the presence or absence of a particular transposon to be monitored. These uses are:

(i)   They can be used to mutate genes and a high proportion of mutants isolated in a strain carrying, for example, Tn10 will have a Tn10 inserted into a bacterial gene. One simple method for isolating Tn10-induced mutations is to: (1) use a temperate phage to infect a strain of *E. coli* harbouring a plasmid with a Tn10 element; (2) upon lysis a few phages (perhaps 10^{-7} to 10^{-8}) will have packaged a fragment of plasmid DNA including Tn10; (3) the phage lysate is used to infect a tetracycline-sensitive strain of *E. coli* at a low MOI (< 0.5) and tetracycline-resistant clones selected on medium containing tetracycline; (4) the tetracycline will kill any uninfected bacteria and any bacteria infected with phages not carrying Tn10, and the only survivors will not only have been infected with a phage carrying Tn10 but the Tn10 element **must** have transposed into the bacterial chromosome; only then can it be replicated and allow the formation of a tetracycline-resistant clone; (5) the tetracycline-resistant clones are now screened for the required mutation (*lac*⁻, for example).

When a transposon is inserted in this way, there is complete linkage between the mutation and the drug-resistance genes, and varying degrees of linkage between the resistance genes and other bacterial genes adjacent to the point of insertion.

(ii) Transposition can greatly facilitate genetic mapping as it is possible (as described above) to place a transposon carrying a selectable drug-resistance marker adjacent to almost any gene on the bacterial chromosome. For example, insertions of Tn10 adjacent to the *lac* operon can be recovered from the tetracycline-resistant clones isolated in the last experiment. Every clone has Tn10 inserted at a different chromosomal location and it is likely that at least one has Tn10 adjacent to *lac*. These can be isolated by: (1) mixing together very many of these clones; (2) growing phage P1 on this family of clones and using the lysate to infect a *lac*⁻ *tet*ˢ recipient; and (3) selecting *lac*⁺ transductants and then testing these for tetracycline resistance. Any *lac*⁺ *tet*ᴿ clones will have arisen by the co-transduction of *lac*⁺ and Tn10 and so should have Tn10 inserted close to the *lac* region.

(iii)   Transposable elements contain a variety of restriction sites, and when a transposon is inserted into a particular gene new restriction sites are introduced into the DNA molecule and restriction enzyme digests will show an altered pattern of fragments. Consequently, it is possible to determine the precise physical location of the transposon (and hence of the gene into which it is inserted) and to correlate the genetic and physical maps.

(iv)   Transposons can be used to provide mobile restriction sites, a property that is sometimes useful in gene cloning.

(v)   Almost any bacterial gene can be mobilised and converted into a transposon by flanking it with two copies of an IS element.

(vi)   IS elements can be used as mobile promoters or terminators of transcription.

(vii)   Transposable elements provide portable regions of homology. This is the basis of Hfr formation (section 3.2.4) but other plasmids can be integrated into the bacterial chromosome, or two plasmids fused together, provided that each replicon carries a copy of the same IS element or composite transposon; these co-integrates are comparatively stable. Similar co-integrates can also form if both replicons carry a complex transposon, such as Tn3, but these co-integrates are very unstable as they are quickly resolved into the component replicons by the efficient site-specific resolution system.

(viii)   IS elements (and some transposons) can be used to generate deletions and other chromosomal rearrangements, facilitating the isolation of new gene arrangements (section 7.5).

Exercises

7.1   What special features distinguish insertion sequences and transposons from each other and from other types of genetic element?

7.2   Describe the importance of the inverted repeat sequences present at the ends of an IS element.

7.3   The composite transposons are flanked by two IS or IS-like elements. Why does it not matter whether these elements are in the same or in opposite orientations?

7.4   Describe two simple methods which would enable you to recognise a plasmid carrying a particular transposon.

7.5   Compare the structure and properties of Tn10 and Tn3.

7.6   Distinguish between replicative and conservative transposition. Why is a short sequence of target site DNA always duplicated during transposition?

7.7   Explain the formation and resolution of co-integrates during Tn3 transposition.

7.8   Conservative transposition produces simple insertions. Distinguish the alternative mechanisms that have been suggested to account for simple insertions.

7.9   How can you account for the occasional co-integrate molecules found in the presence of certain IS elements and class I transposons?

7.10   Explain how transposition sometimes results in the deletion or inversion of an adjacent segment of DNA.

7.11   A particular IS element has been inserted at several different locations within the *E. coli galE* gene. When the element is in one orientation neither *galE* nor the downstream *galT* and *galK* genes are expressed, but in the other orientation only *galE* is non-functional. Explain.

7.12   You have constructed a new plasmid which can replicate at $25°C$ but not at $42°C$. This plasmid, which also carries a copy of Tn10, was introduced into a tetracycline-sensitive strain of *E. coli* and a tetracycline-resistant clone isolated, grown in nutrient broth and plated on medium containing tetracycline (about 10^8 bacteria per plate). When the plates were incubated at $42°C$, a few tetracycline-resistant and non-temperature-sensitive colonies were recovered.

What alternative mechanisms could account for these colonies? What further experiments could help you to decide which particular mechanism(s) was involved?

References and related reading

Bennet, P., 'Bacterial transposons', in *Bacterial Genetics* (eds J. Scaife, D. Leach and A. Galizzi), Academic Press, London, p. 17 (1985).

Cohen, S. W. and Shapiro, J. A., 'Transposable genetic elements', *Scientific American*, (Feb.) **242**, 40 (1980).

Cullum, J., 'Insertion sequences', in *Bacterial Genetics* (eds J. Scaife, D. Leach and A. Galizzi), Academic Press, London, p. 85 (1985).

Grindley, N. D. F. and Reed, R. R., 'Transpositional recombination in prokaryotes', *Ann. Rev. Biochem.*, **54**, 863 (1985).

Heffron, F., 'Tn3 and its relatives', in *Mobile Genetic Elements* (ed. J. Shapiro), Academic Press, New York, p. 223 (1983).

Iida, S., Meyer, J. and Arber, W., 'Prokaryotic IS elements', in *Mobile Genetic Elements* (ed. J. Shapiro), Academic Press, New York, p. 159 (1983).

Kleckner, N., 'Transposable elements in prokaryotes', *Ann. Rev. Genetics*, **15**, 341 (1981).

Shapiro, J. A., 'Molecular model for the transposition and replication of bacteriophage Mu and other transposable elements', *Proc. Natl. Acad. Sci. USA*, **76**, 1933 (1979).

Starlinger, P., 'DNA rearrangements in procaryotes', *Ann. Rev. Genetics*, **11**, 103 (1977).

BACTERIOPHAGES 8

8.1 Introduction

Phages are invaluable tools to both the geneticist and the molecular biologist. Their use has led to the discovery of many fundamental principles of molecular genetics and has had a critical impact on our thinking in these areas. Much of our knowledge of DNA replication came initially from studies with ϕX174 while studies with λ have revealed the highly complex network of systems which regulates the expression of the λ genes — systems with many hitherto unknown features. More recently, several phages have been developed as cloning vectors.

In this chapter we introduce three new phages, Mu, X174 and M13, which have been extensively used in molecular studies, and examine a further aspect of λ, the mechanism of site-specific recombination.

8.2 Bacteriophage Mu

The structure of Mu is very similar to that of λ and it has a polyhedral head (60 nm in diameter), a tail with a contractile sheath (60–100 × 20 nm) and a base plate to which are attached several tail fibres; within the head is a single linear molecule of double-stranded DNA.

Mu, discovered in 1963 by **Austin Taylor**, is a most unusual phage and has proven to be a most useful tool in both bacterial and molecular genetics; it is both a virus and a transposable element (it was the first bacterial transposable element to be discovered) and transposition is an obligatory part of its life style. Like λ, Mu is a temperate phage and it can either lyse or lysogenise an infected cell, but there any similarity ends.

8.2.1 The Mu genome

The DNA molecules extracted from mature Mu virions are linear and about 39 kb long. Each molecule is made up of 37.5 kb of non-permuted

Mu DNA flanked by heterogeneous sequences of DNA derived at random from the previous host, 50–150 bp at the left and about 1.5 kb at the right (figure 8.1); thus, in effect, the Mu genome is an integrated prophage.

The Mu genome itself is divided into 3 regions:

(1)   The α region, 33 kb long and including most of the 32 identified genes. From our point of view the most important genes are
    c encoding the Mu repressor which can bind to an operator site just to the right of *c* and repress most of the other Mu genes. *c* is the equivalent of the *cI* gene in lambda
    A encoding the Mu transposase
    B encoding a replication protein
    Two groups of genes encoding the proteins required for head and tail assembly.

(2)   The 3 kb G-segment, an invertible region containing genes encoding host range specificity.

(3)   The 1.7 kb β segment. This contains only two genes one of which, *gin* (*G in*version), is responsible for inversion of the G segment (section 8.2.4).

Note that Mu does not have cohesive ends nor is the genome terminally redundant (that is, the sequence at one end of the genome is not directly repeated at the other end). This means that the Mu genome cannot circularise and, consequently, cannot integrate into the host chromosome in the same way as λ (section 1.5.2).

Figure 8.1
The structure and integration of phage Mu

Infective Mu DNA (above) is flanked by short sequences derived from the previous host. When Mu infects and lysogenises a new host cell, it integrates into the *E. coli* chromosome by undergoing conservative transposition (centre; see also figure 7.7); the sequences from the previous host are lost and a 5 bp target sequence of new host DNA is duplicated (below).

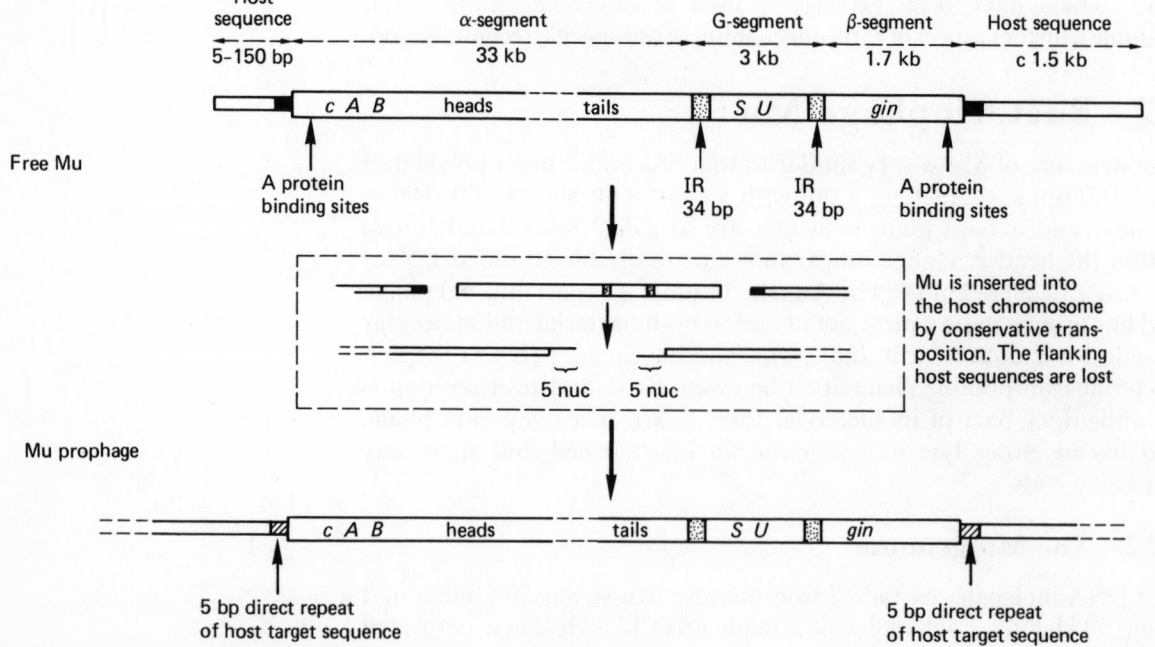

8.2.2   Lysogenisation by Mu

Mu integrates into the host chromosome by **transposing** from the
39 kb molecule of virion DNA (that is, the Mu DNA itself together
with the flanking sequences of host DNA) into the host chromosome.
This is thought to be a conservative process (see sections 7.4 and 7.4.2)
and it results in the loss of the heterogeneous sequences of flanking host
DNA and generates a 5 bp direct repeat of the target site DNA
(figure 8.1).

When Mu integrates in this way it appears to do so more or less
at random — it behaves just like all the other transposable elements.
Mu has been shown to integrate into the *lacZ* gene at at least 60 different
sites, sometimes in one orientation and sometimes in the other, and
because of this lack of integration specificity it can also integrate into
plasmids. Lysogenisation by Mu is a rather inefficient process and some
90 per cent of the cells surviving Mu infection are non-lysogenic sensitive
cells. However, several per cent of the lysogens have a mutant phenotype
and in most instances there is complete linkage between the mutation
and the integrated prophage; these mutations occur as a result of Mu
integrating into a structural gene and inactivating it, so that Mu acts
as a **mutator** phage (hence its name, **mu**tator). Other lysogens have
a segment of DNA deleted adjacent to the prophage; these are adjacent
deletions corresponding to those induced by IS1 (section 7.5).

8.2.3   Mu – the lytic cycle

When Mu infects a sensitive cell it must, irrespective of whether it
follows the lytic or lysogenic pathway, first integrate into the DNA of
the host cell; this, as described above, occurs by conservative
transposition. In most infected cells, Mu enters the lytic cycle and
commences to replicate; this is a unique process and occurs by Mu
undergoing successive cycles of replicative transposition (section 7.4.1)
until there are 50–100 copies of the Mu genome present, each copy
inserted into a different region of the host DNA. This means that up
to 100 transpositions occur in each Mu-infected cell. During this period
of replication the Mu late genes are expressed and the virion proteins
are synthesised and assembled. Eventually the Mu genomes are
packaged into the phage heads by a **headful packaging** mechanism.
Packing commences within the flanking host DNA at the *c* end of the
genome and occurs in such a way that between 50 and 150 bp of this
adjacent host DNA is packaged, and the process then continues until
the phage heads are 'full'. A headful is about 39 kb, so that about 1.5
kb of the host DNA at the *β*-end of the genome is also packed; if a
deletion is present in the Mu DNA, then the Mu genome is effectively
reduced in size and this is compensated by increasing the length of the
host DNA at the *β*-region end. The mechanism that generates the
variable ends at the *c* end of the genome is not understood, but it is
curious that the attached host sequences are all multiples of 11 bp —
the number of base pairs in one complete turn of the double helix.

A feature of Mu replication is the very high frequency of

chromosomal aberrations that is induced by the transposition process. These abnormal products cannot be recovered from normally infected cells as the host DNA molecules are destroyed when the Mu DNA is packaged, but they can be recovered from cells infected by Mu mutants which can replicate their DNA but are unable to complete the later stages of the lytic cycle. Among these aberrant products are found co-integrates (a normal product of replicative transposition), a variety of circular molecules of various sizes each containing a single copy of Mu (several per cell), and molecules with either deletions (10^{-2} per cell) or inversions (several per cell).

The transposition of Mu requires both ends of Mu and the A gene product and, if either end of Mu is missing or if the A gene is non-functional, then both integrative transposition (lysogenisation) and replicative transposition (Mu replication) are abolished. Both types of transposition are stimulated by the B gene product, and B^- mutants have greatly reduced abilities to lysogenise and to replicate.

Unlike other transposable elements, Mu does not have terminal inverted repeat sequences that can be recognised by the transposase. Instead there are three 11-bp sequences near each end of Mu which are binding sites for the A gene protein, the transposase. One sequence is at positions -17 to -27 from each Mu end, and the other two copies are within 175 nucleotides of the left end and 80 nucleotides of the right end, and it is probable that the transposase promotes transposition by first binding to one or more of these sequences at each end of the Mu genome.

8.2.4 The invertible G-segment of Mu

Another unusual feature of Mu is the invertible G-segment and, depending on the orientation of this segment, Mu can infect either *E. coli* K12 (the G(+) orientation) or *E. coli* C (the G(−) orientation). This change in host range specificity is achieved by switching on the alternative sets of genes $S U$ [G(+)] and $S' U'$ [G(−)]; these genes are involved in Mu adsorption and their products are probably components of the tail fibres.

The probable organisation of the G-segment is shown in figure 8.2. Note that part of the S gene, the S_C or constant region, lies outside the G-segment and is common to both S and S'; the remaining part of the S gene, S_V or S'_V is variable and depends upon the orientation of the G-segment. As a result, the S and S' proteins have the same amino acid sequence at the amino termini but different sequences at the COOH ends. Both S and U (and S' and U') are transcribed from a promoter located to the left of S_C.

The segment is flanked by a pair of 34 bp inverted repeats, and inversion is the result of a site-specific recombination event occurring between these repeat sequences; in just the same way that site-specific recombination in λ requires the integrase protein, so site-specific recombination in Mu requires the product of the *gin* (*G-in*version) gene; *gin* lies just outside the G-segment in the adjacent β-region.

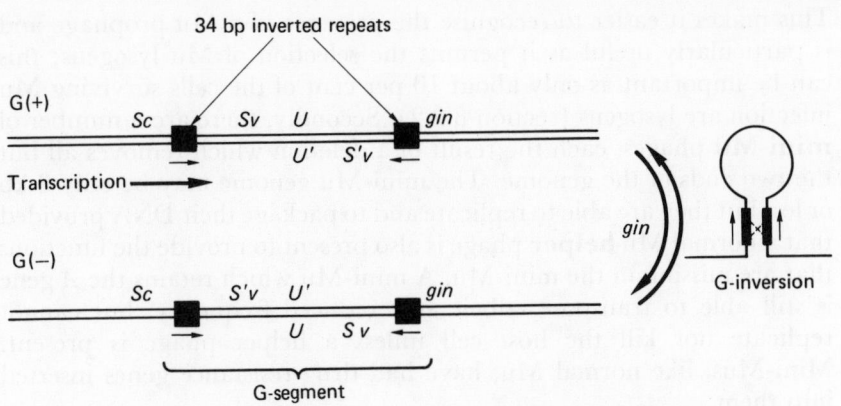

Figure 8.2
The invertible G-segment

The 3 kb G-segment is inverted by reciprocal site-specific recombination occurring between the flanking 34 bp inverted repeats (right); this requires the activity of the *gin* encoded recombinase.

In the G(+) orientation, *S* and *U* are transcribed and Mu can only infect *E. coli* K12; in the G(−) orientation, *S'* and *U'* are transcribed and Mu can only infect *E. coli* C. Note that the left end of the *S* gene (*Sc*) lies outside the G-segment and is common to both *S* and *S'*.

Inversion of the G-segment is a comparatively rare event and only occurs about once in every three generations. This is because the *gin* protein is only present in very small amounts and, furthermore, because *gin* is only expressed in lysogenic cells, G-inversion **only** takes place in Mu lysogens. As a result of G-inversion, in a large population of lysogenic cells about 50 per cent of the cells will have a Mu prophage with the G-segment in the G(+) orientation while the remaining 50 per cent will have a Mu G(−) prophage. When this mixed population of lysogenic cells is induced, the Mu G(+) and Mu G(−) lysogens will produce only Mu G(+) and only Mu G(−) progeny respectively; if this lysate is now used to infect *E. coli* K12, only the Mu G(+) phages (about one-half of the phages present in the lysate) can infect the host cells and, in turn, this will produce Mu G(+) lysogens and Mu G(+) progeny (via the lytic response). It is only after a new lysogen has been grown for many generations that G-inversion will have produced a mixed population of Mu G(+) and Mu G(−) lysogens.

8.2.5   The genetic uses of Mu

Because Mu shows the properties of both a virus and a transposon, it has great versatility as a genetic tool. On the one hand, as a temperate phage it can either lyse or lysogenise the cell and, like P1, can carry out general transduction as a result of the rare packaging of bacterial DNA into phage heads. On the other hand, Mu behaves as a transposable element; lysogenisation occurs by the conservative transposition of Mu into the chromosome or a plasmid within the infected cell, while replication occurs by the repeated replicative transposition of Mu to random sites in the host's genome.

Two types of Mu have been particularly useful in molecular genetic experiments. Firstly, Mu variants with a mutation in the *c* (repressor) gene which results in the production of a temperature-sensitive repressor protein; these Mu *cts* mutants have the advantage that they are easily induced by heating to 42°C, at which temperature the repressor is reversibly inactivated. A further refinement has been the introduction of an ampicillin or kanamycin resistance gene into the Mu genome.

This makes it easier to recognise the presence of a Mu prophage and is particularly useful as it permits the selection of Mu lysogens; this can be important as only about 10 per cent of the cells surviving Mu infection are lysogens (section 8.2.2). Secondly, there are a number of **mini-Mu** phages, each the result of a deletion which removes all but the two ends of the genome. The mini-Mu genome may be only 2 kb or less but they are able to replicate and to package their DNA provided that a normal Mu **helper** phage is also present to provide the functions that are missing in the mini-Mu. A mini-Mu which retains the *A* gene is still able to transpose, albeit at a reduced frequency, but cannot replicate nor kill the host cell unless a helper phage is present. Mini-Mus, like normal Mu, have had drug-resistance genes inserted into them.

Some of the special uses of Mu (and mini-Mu) are as follows:

(i)   Mu is a mutator phage and can be used to induce mutations. Between 1 and 3 per cent of all lysogens show a mutant phenotype and, for any particular gene, mutations due to the insertion of Mu are up to 10 times more frequent than spontaneous mutations.

(ii)   Mu induces deletions, inversions and, probably, duplications.

(iii)   Mu can cause replicon fusion (co-integrate formation). If, for example, Mu is carried on a plasmid, then it can fuse that plasmid either to another plasmid or to the *E. coli* chromosome. This is the consequence of intermolecular transposition and is similar to the formation of co-integrate molecules during the transposition of Tn3 (figures 7.4 and 7.5).

(iv)   In 1976, **Michel Faelen** and **Ariane Toussaint** showed that when a cell of *E. coli* contains both an autonomous plasmid (such as F'_{lac}) together with a copy of Mu on the bacterial chromosome, then the Mu prophage can stimulate the transposition of particular genes from the bacterial chromosome to the plasmid. This is a comparatively rare event (10^{-4}) and the transposed DNA is inserted at random into the plasmid and is flanked by two copies of Mu in the same orientation. The segment of transposed DNA is between 9 and 150 kb long (0.25 to 3.5 minutes on the *E. coli* map) and, if two genes fall within the same segment, they can be **co-transposed**. Furthermore, as the distance between two genes decreases so the frequency of co-transposition increases, making it possible to order closely linked genes on the bacterial chromosome. This process not only provides a further method for mapping the bacterial chromosome but also, because almost any chromosomal gene can be transferred to a plasmid, facilitates the construction of F' and R plasmids carrying particular bacterial genes.

These DNA rearrangements are all consequences of Mu transposition and so can only occur when the *A* gene protein is present. Since the *A* gene is not transcribed in lysogens, these events can only be observed after Mu induction. Mu also provides regions of homology between which general recombination can occur — thus even in the absence of a functional transpositional system it is still possible to detect a wide variety of DNA rearrangements provided that a functional Rec$^+$ system is present.

8.3   Site-specific recombination in λ

Several examples of site-specific recombination have been described — G-segment inversion in Mu (section 8.2.4), the resolution of Tn3 transpositional co-integrates (section 7.4.2) and the integration and excision from the *E. coli* chromosome of bacteriophage λ (section 1.5.2). This process differs from general recombination in two important ways; firstly, it is *recA* independent and instead requires a special site-specific recombinase together with a host-cell-encoded protein, the **integrative host factor** (IHF); secondly, site-specific recombination requires only very limited homology between the two molecules involved and it occurs between highly specific DNA sequences rather than between two molecules with extensive sequence homology.

The first and best-known example of site-specific recombination is the integration and excision of λ which has been studied both at a genetic level and, more recently, at a molecular level by **Arthur Landy and his colleagues**. The integration of λ requires site-specific recombination between specific attachment sites (see figure 1.13) on the phage and bacterial chromosomes (*attP* or *PoP'* and *attB* or *BoB'*) and generates hybrid attachment sites at each end of the integrated prophage, *attL* or *BoP'* and *attR* or *PoB'*); this event requires the λ *int* gene product and the IHF protein. Excision (figure 1.14) is the reverse process and the result of site-specific recombination between *attL* and *attR* regenerating the original host and phage chromosomes with their respective attachment sites; this reaction also requires the product of the λ *xis* gene.

Site-specific recombination occurs between these specific attachment sites because the *int, xis* and IHF proteins recognise and bind to particular sequences within these sites.

8.3.1   The structure of the attachment sites

Each *att* site consists of a 15 bp A–T-rich common core (*o*) sequence within which crossing-over occurs, and a pair of flanking **arms**, each with its own nucleotide sequence. The function of each one of these arms (*P*, *P'*, *B* and *B'*) has been studied using a hierarchical series of deletion mutants where the different deletions all originate within the adjacent phage or bacterial DNA and extend progressively further into the arm; thus, in effect each deletion removes further nucleotides from the outer extremity of the arm. These investigations have shown that whereas *attP* cannot function if it is less than 240 bp long (figure 8.3), *attB* continues to function provided that the 4 bp on each side of the core sequence are present. This means that *attP* and *attB* are functionally different and that *attP* must play a more complex role than *attB*.

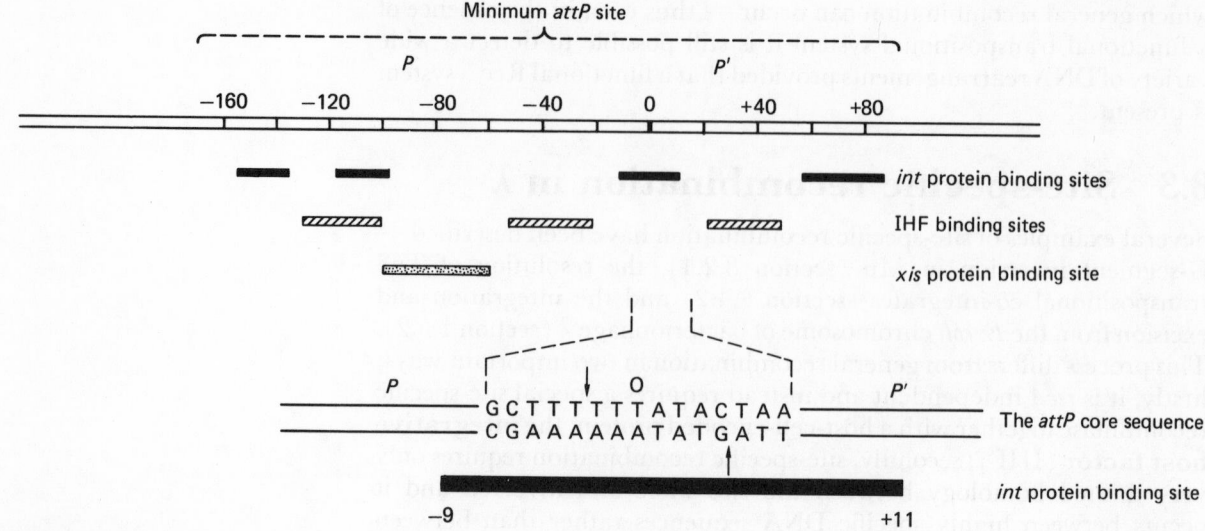

Figure 8.3
The structure of *attP*

The λ *attP* has a 15 bp core sequence (common to all *att* sites) flanked by a pair of dissimilar arms, *P* and *P′*. This region has specific binding sites for the λ *int* and *xis* proteins and the host IHF. The arrows show the position of the staggered nicks made by the *int* protein to initiate site-specific recombination.

The bacterial *attB* site has the same core sequence but the flanking arms, *B* and *B′*, are only 4 bp long; thus *attB* can only bind the *int* protein. The nucleotides are numbered from the central base in the core. (Adapted from Weisberg, R. A. and Landy, A., in *Lambda II* (eds R. W. Hendrix *et al.*), Cold Spring Harbor Laboratory, New York, p. 211 (1983).)

The three proteins involved in site-specific recombination, the *int* and *xis* gene products and the IHF protein, all bind to different sites within *attP*. The integrase protein binds to one 20 bp sequence within the core region and another within *P′* and to two shorter sequences within *P*; this shows that the integrase protein must have two different binding specificities. The IHF binds to three sequences between 22 and 31 bp long and the *xis* protein binds to a single 40 bp sequence within *P* (figure 8.3). On the other hand, *attB* consists primarily of a core sequence and so only binds the *int* protein.

8.3.2   *attP* × *attB* site-specific recombination

Although a considerable amount is known about the interactions between these proteins and the DNA at the *att* sites, very little is known about the molecular basis of site-specific recombination. Recombination is initiated by a pair of staggered single-strand nicks made exactly 7 bp apart in the opposite strands of each participating duplex and, since the recombinant *att* sites always have exactly the same core sequence, it is supposed that these nicks are always made between the same nucleotides. It then seems that the participating molecules separate and rejoin in a new combination as the result of strand separation and strand switching. How this occurs is not known but the outcome is shown in figure 8.4.

The other systems of site-specific recombination are not as well known but it is interesting to note that the inverted repeats flanking the G-segment of Mu and the H-segment in *Salmonella typhimurium* show considerable sequence homology and that the *gin* and *hin* gene products are functionally interchangeable. Furthermore, the Tn3 resolvase, although not functionally interchangeable with the *gin* and *hin* proteins, does closely resemble them in its amino acid sequence. One suggestion is that all these systems are descended from a common ancestral system.

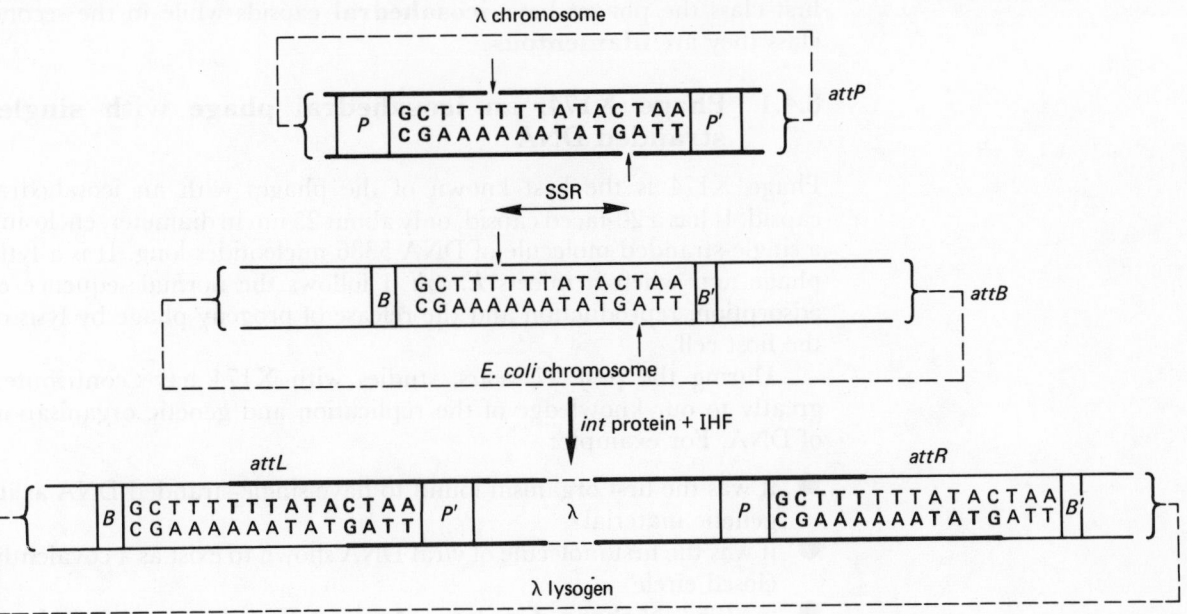

Figure 8.4
**Site-specific recombination
between *attP* and *attB***

The λ *int* protein makes pairs of
staggered nicks, 7 bp apart, within the
attP and *attB* core sequences (top),
and an unknown mechanism of strand
exchange joins the right end of the
attP core to the left end of the *attB*
and vice versa. Note that
recombination is reciprocal and that,
although 7 bp segments of
heteroduplex DNA are formed, there
are no mismatched base pairs; this is
because all the cores normally have
the same nucleotide sequence.

8.4   Phages with single-stranded DNA

Not all DNA is double-stranded and some of the smallest phages have
a circular molecule of single-stranded DNA as their genetic material.
This was first inferred by **Robert Sinsheimer** in 1959 as a result of his
studies with the DNA of phage X174. This DNA had three unusual
properties; firstly, it did not have equivalent amounts of adenine (25
per cent) and thymine (33 per cent), nor of cytosine (24 per cent)
and guanine (18 per cent), so that it was impossible for all the bases
to be in the form of conventional base pairs; secondly, the DNA was
very much more sensitive to the action of formaldehyde than was
normal double-stranded DNA; and, thirdly, unlike other DNA, it did
not behave like rigid rods when in solution.

These phages are characterised by an unusual pattern of replication.
This takes place in three stages:

(i)   The single strand of viral (+) DNA is converted into a
       double-stranded **replicative form** (RF) by the synthesis of
       the complementary (−) strand.
(ii)   The double-stranded replicative form undergoes several cycles
       of **rolling-circle replication** (section 1.5.1) to produce
       further molecules of replicative form DNA.
(iii) When sufficient molecules of RF DNA have accumulated,
       rolling-circle replication switches to producing single-stranded
       viral (+) genomes; these circularise and are eventually
       packaged into phage capsids.

There are two classes of phages with single-stranded DNA. In the first class the phages have **icosahedral** capsids while in the second class they are **filamentous**.

8.4.1 Phage X174, an icosahedral phage with single-stranded DNA

Phage X174 is the best known of the phages with an icosahedral capsid. It has a 20-faced capsid, only about 25 nm in diameter, enclosing a single-stranded molecule of DNA 5386 nucleotides long. It is a lytic phage and when it infects *E. coli* it follows the normal sequence of adsorption, reproduction and the release of progeny phage by lysis of the host cell.

During the past 27 years, studies with X174 have contributed greatly to our knowledge of the replication and genetic organisation of DNA. For example:

- it was the first organism found to have single-stranded DNA as its genetic material
- it was the first molecule of viral DNA shown to exist as a covalently closed circle
- in 1967 **Mehran Goulian, Arthur Kornberg and Robert Sinsheimer** copied X174 DNA in an *in vitro* system and made the first biologically active (that is, infective) DNA
- in 1968 **David Dressler and Walter Gilbert** first proposed the rolling-circle model for replication to explain the production of viral (+) strands from the double-stranded replicative form DNA
- in 1977 **Fred Sanger and his colleagues** established the complete nucleotide sequence of X174 DNA; this was the first DNA genome to be completely sequenced

One of the most remarkable features to emerge from the sequencing of the genome was the confirmation of the concept of **overlapping** genes. The first hint that the genes of X174 might overlap came from mapping experiments, when it was found that some mutations within the B gene mapped in between two different A gene mutations. Furthermore, it was known that the 10 proteins encoded by X174 totalled 1986 amino acid residues; this requires at least 5958 nucleotides, whereas the actual genome was only 5386 nucleotides long. It turns out that only three of the 11 genes of X174 are free from overlaps; not only do the termini of some genes overlap the starts of others, but the *B* gene is located entirely within *A, E* is within *D*, and *K* is partly within *A* and partly within *C* (figure 8.5). When genes overlap in this way they are translated in different reading frames, so that the first nucleotide in the codon of one gene will either be the second or third nucleotide in the codon of an overlapping gene.

8.4.2 M13, A filamentous phage with single-stranded DNA

The filamentous phage M13, a member of the second class of single-stranded DNA phages, has a very different life style and is of

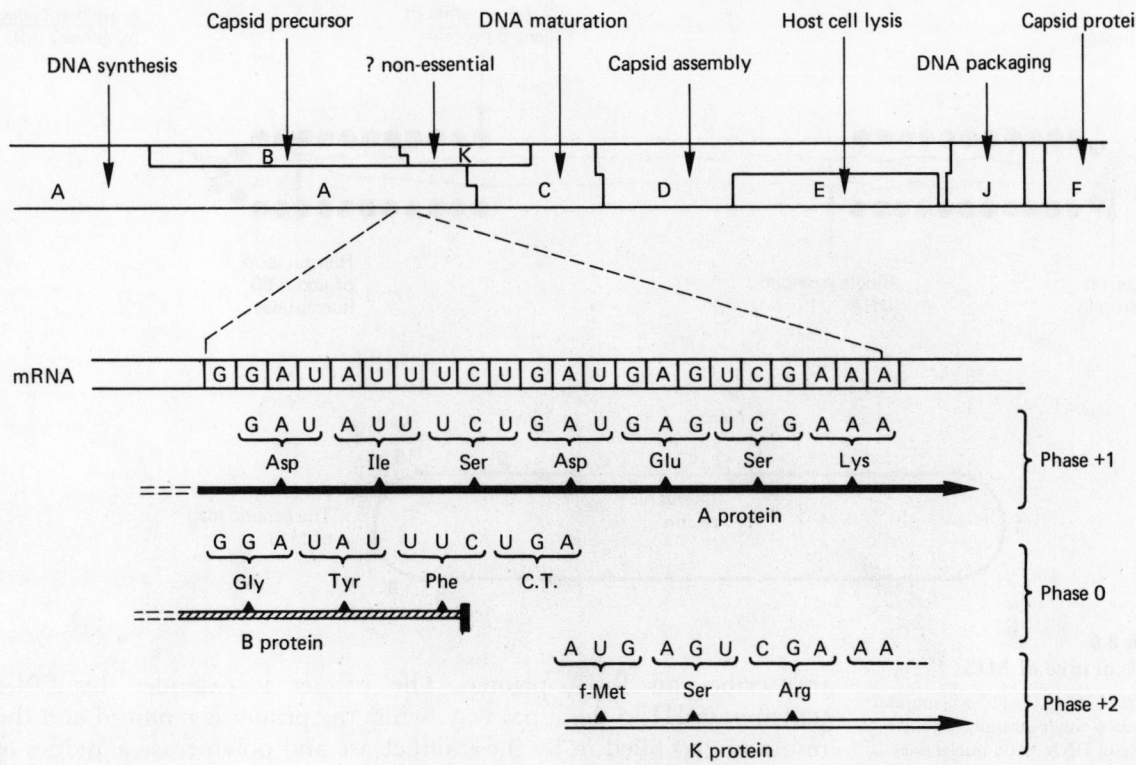

Figure 8.5
Overlapping genes in phage X174

The segment shown at the top is about 1.5 kb long (about 25 per cent of the X174 chromosome) and each of the genes, *A, B, C, D, E, J,* and *K* overlaps at least one other gene. The lower part of the figure shows the nucleotide sequence of the messenger RNA spanning the overlap between the end of *B* and the start of *K. B* and *K* overlap by one nucleotide and they are read in different phases of the genetic code, different from each other and from the overlapping *A* gene. (C. T. indicates a chain termination (nonsense) codon.)

importance as it is increasingly used as a vector for cloning DNA, particularly when the cloned genes are required in the form of single-stranded DNA. The genome is circular, 6408 nucleotides long (about 2 μm) and, unlike X174, does not include any conspicuous overlapping sequences.

The capsid of M13 is 895 × 9 nm and the 'cylinder' that makes up the filament is assembled from 2700 copies of the gene 8 protein. On one end is a group of protein molecules encoded by genes 7 and 9 while the other end is assembled from molecules of the gene 6 protein, to which are attached knobs of the gene 3 protein (figure 8.6).

When M13 infects an *E. coli* cell it adsorbs to the tip of an F-pilus; adsorption cannot occur in the absence of an F-pilus (M13 is another **male-specific** phage) nor if the knobs encoded by gene 3 are missing. The relatively intact phage particle is now thought to enter the host cell and the phage capsid is, at least in part, degraded within the host cell and the breakdown products are used in the assembly of capsids for the progeny phage.

The single-stranded molecule of (+) DNA released into the host cell is, with the exception of a tract of about 60 nucleotides, protected by the formation of a hairpin loop, immediately coated with the host-cell-encoded single-strand binding protein (figure 8.7). The uncoated hairpin loop acts as a replication origin and is recognised by a molecule of RNA polymerase, which opens up the hairpin and

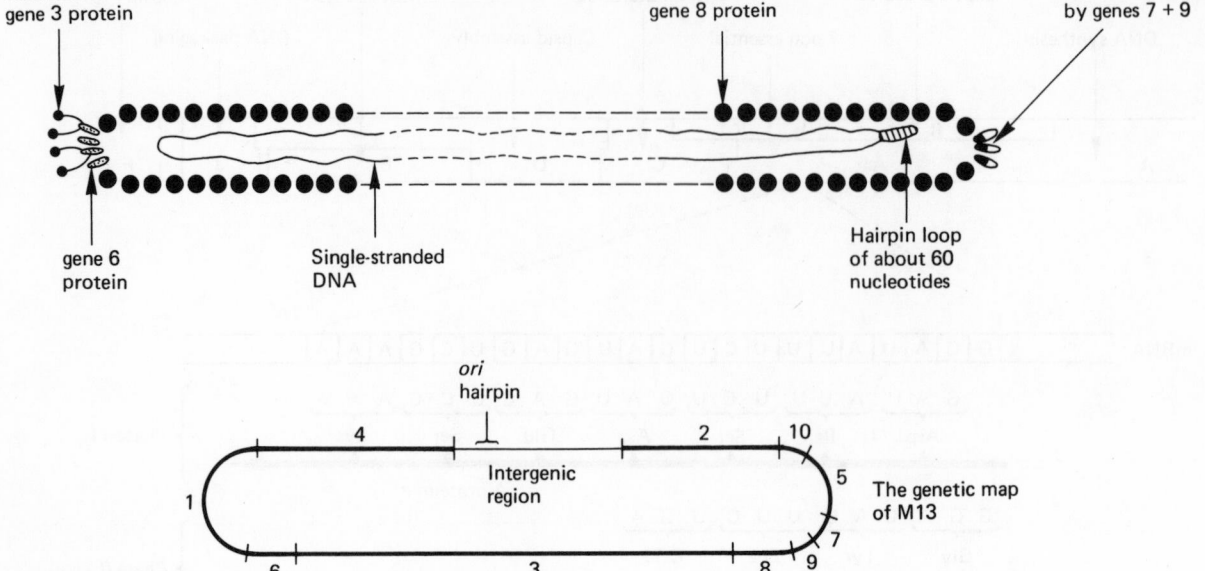

Figure 8.6
The structure of M13

The capsid of M13 is 895 × 9 nm and
it encloses a single-stranded circular
molecule of DNA, 6408 nucleotides
long. The lower figure shows the
position of the 10 genes of M13 on the
chromosome.

transcribes an RNA primer. The primer is extended by DNA
polymerase III in the usual way, while the primer is removed and the
resulting gap filled in by the exonuclease and polymerase activities of
(probably) DNA polymerase I; finally the two free ends of the new
strand of DNA are ligated to form a covalently closed circular molecule
of double-stranded replicative form (RF) DNA.

In addition to the replication system of the host cell, the conversion
of single-stranded to replicative form DNA also requires the gene 3
protein; this is one of the proteins making up the phage capsid and it
appears to associate with the hairpin loop and in some way promote
the production of RF DNA. Both the remaining steps involve rolling-
circle replication and require the phage gene 2 product, an endonuclease
which initiates each replication cycle by nicking the (+) strand of the
RF DNA at a specific replication origin. The molecule of RF DNA
now undergoes several cycles of rolling-circle replication and this
continues until about 100 molecules of double-stranded daughter RF
DNA are present; these molecules act both as templates for transcription
and for the production of single-stranded viral (+) DNA.

The final step is the switching of rolling-circle replication to the
production of single-stranded viral DNA, and this occurs when
transcription and translation of the RF DNA has produced sufficient
gene 5 protein. This protein is found associated with the single-stranded
linear molecules cut off from the concatamer produced by rolling-circle
replication, converting them to a distinctive nucleoprotein; it seems
likely that one of the roles of the gene 5 protein is to prevent the
synthesis of a complementary (−) strand. Finally, the linear molecules
circularise and new phage capsids assemble around them, at the same
time displacing the phage gene 5 protein.

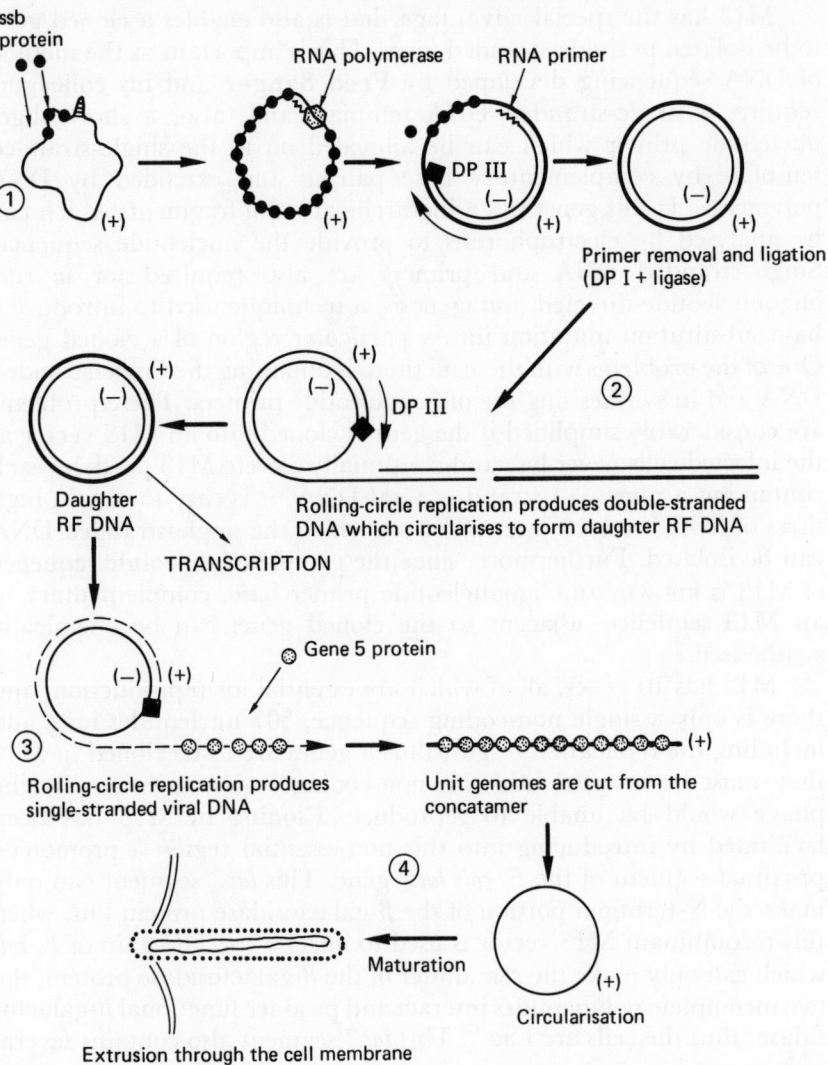

Figure 8.7
The replication of M13

(1) The conversion of single-stranded
(+) DNA to the double-stranded
replicative form.
(2) Rolling-circle replication of the
replicative form.
(3) When sufficient gene 5 protein
has accumulated, rolling-circle
replication produces single strands of
viral (+) DNA.
(4) After circularising, the single
strands are packaged into capsids and
the mature phages are extruded
through the cell membrane.

As the phage particles are assembled they are released from the
cell by being extruded through the cell membrane. This process is a
continuous one and some 1000 phage particles are released from the
host cell each generation. Observe that the M13 genome is never
integrated into the host cell chromosome, it never becomes prophage
and it never becomes quiescent, and yet the host cell is never lysed.

There are normally some 100 copies of RF DNA in an infected
cell and, whenever the cell divides, each daughter cell will receive
multiple copies of this DNA; this RF DNA will commence to replicate
and will continue to do so until once again there are about 100 copies
present. This double-stranded RF DNA behaves in much the same
way as a high copy number plasmid so that M13 replicative form DNA
can be isolated by the methods used to isolate plasmid DNA.

M13 has the special advantage that is also enables a cloned gene to be isolated in single-stranded form. This is important as the method of DNA sequencing developed by **Fred Sanger** and his colleagues requires a single-stranded DNA template and, also, a short oligonucleotide primer which can be annealed on to the single-stranded template by complementary base pairing and extended by DNA polymerase I; this generates a hierarchical set of fragments which can be analysed by electrophoresis to provide the nucleotide sequence. Single-stranded DNA and primers are also required for *in vitro* oligonucleotide-directed mutagenesis, a technique used to introduce a base substitution mutation into a particular region of a cloned gene. One of the problems with these methods is obtaining the single-stranded DNA and in synthesising the oligonucleotide primers. These problems are considerably simplified if the gene is cloned into an M13 vector as the infected cells never lyse and continually secrete M13 particles each containing a plus (+) strand of viral DNA; it is easy to obtain high titres of M13 (over 10^{12} per ml) from which the single-stranded DNA can be isolated. Furthermore, since the complete nucleotide sequence of M13 is known, an oligonucleotide primer base, complementary to an M13 sequence adjacent to the cloned gene, can be chemically synthesised.

M13 has 10 genes, all of which are essential for reproduction, and there is only a single non-coding sequence, 507 nucleotides long and including the replication origin; thus if genes are to be cloned in M13 they must be inserted into this non-coding region as otherwise the phage would be unable to reproduce. Cloning in M13 has been facilitated by introducing into this non-essential region a promoter-proximal segment of the *E. coli lacZ* gene. This *lacZ* segment can only make the N-terminal portion of the β-galactosidase protein but, when this recombinant M13 vector is used to infect a special strain of *E. coli* which can only make the remainder of the β-galactosidase protein, the two incomplete polypeptides interact and produce functional β-galactosidase; thus the cells are Lac $^+$. This *lacZ* segment also contains several unique restriction sites (that is, sites not present elsewhere on the vector molecule) and the insertion of a foreign gene into one of these sites results in the loss of β-galactosidase activity and in expression of the Lac $^-$ phenotype.

Exercises

8.1   Describe how Mu lysogenises an *E. coli* host cell; contrast this with lysogenisation by λ.

8.2   How does Mu replicate? Which genes are essential for replication?

8.3   Mu has been described as a giant transposon masquerading under the cloak of a virus. Discuss this statement.

8.4   How can you account for the heterogeneous sequences of host DNA found at each end of the free Mu genome?

8.5   Describe G-inversion in Mu. Is this related to other systems of site-specific recombination?

8.6 Compare and contrast site-specific and general (homologous) recombination. What are the roles of the *int* and *xis* gene products in promoting site-specific recombination in λ?

8.7 List the differences and similarities between the replication of λ and of M13 DNA.

8.8 Mu is a transposable element but it does not have inverted repeat sequences at the ends of its genome. How then does the transposase recognise the ends of Mu and so promote transposition?

8.9 What are the advantages of M13 as a cloning vector?

References and related reading

Mizuuchi, K., Mizuuchi, M. and Craigie, R., 'The mechanism of transposition of bacteriophage Mu', *Cold Spring Harbor Symp. Quant. Biol.*, **49**, 835 (1984).

Silverman, M. and Simon, M., 'Phase variation and related systems', in *Mobile Genetic Elements* (ed. J. Shapiro), Academic Press, New York, p. 537 (1983).

Smith, G. R., 'Site-specific recombination', in *Bacterial Genetics* (eds J. Scaife, D. Leach and A. Galizzi), Academic Press, London, p. 147 (1985).

Symonds, N., Toussaint, A., Van de Putte, P. and Howe, M. M. (eds), *Phage Mu*, Cold Spring Harbor Laboratory, New York (1987).

Toussaint, A., 'Bacteriophage Mu and its use as a genetic tool', in *Bacterial Genetics* (eds J. Scaife, D. Leach and A. Galizzi), Academic Press, London, p. 115 (1985)

Toussaint, A. and Resibois, A., 'Phage Mu: transposition as a life style', in *Mobile Genetic Elements* (ed. J. Shapiro), Academic Press, New York, p. 105 (1983).

Weisberg, R. and Landy, A., 'Site-specific recombination in phage lambda', in *Lambda II* (eds R. W. Hendrix, J. W. Roberts, F. W. Stahl and R. Weisberg), Cold Spring Harbor Laboratory, New York, p. 211 (1983).

GLOSSARY

Abortive transduction An event where the fragment of DNA introduced by transduction fails to be recombined into the recipient chromosome and is inherited unilinearly.

Adenosine triphosphate (ATP) A compound with high energy phosphate bonds that provide the energy for many cellular processes.

Alleles Alternative forms of the same gene (for example, $lacZ1$, $lacZ2$ and $lacZ^+$).

AP endonuclease An endonuclease which recognises an AP site and cuts the defective strand on the 5′ side of the missing base.

Apurinic or apyrimidinic (AP) site A molecule of single-stranded or double-stranded DNA with a missing purine or pyrimidine base.

Attachment site (att) The specific sequences on phage and bacterial chromosomes between which site-specific recombination occurs in order to integrate the phage genome into the bacterial chromosome.

Autoradiography A photographic image is produced by labelling a molecule, such as a bacterial or phage chromosome, with a radioactive label and overlaying it with a photographic emulsion. The emissions from the radioactive label expose the emulsion and, after development, reveal the size and shape of the underlying molecule.

Auxotroph A mutant micro-organism that will only grow when a particular amino acid, nucleotide or vitamin is provided.

Back-mutation A mutation which results in a mutant gene regaining its normal activity by restoring the exact nucleotide sequence present in the wild type.

Cairns-type replication *see* **Theta-type replication.**

Carcinogen A physical or chemical agent that causes cancer.

Chain-termination (nonsense) triplet (CTT) A codon signalling the termination of Polypeptide synthesis; they normally occur at the ends of genes but may be generated within genes by mutation. These triplets are UAG (amber), UAA (ochre) and UGA (opal).

Chi-form   A recombinational intermediate.

Chi site   An 8-nucleotide sequence on a molecule of DNA, particularly on the λ chromosome, which is a hot-spot for general recombination.

Chiasma   The crossed-over strands of two non-sister chromatids seen at the first meiotic division of eucaryotes; the position at which two homologous chromosomes appear to exchange genetic material during meiosis.

Chimeric plasmid   A plasmid used in gene cloning and constructed from two or more different plasmids.

Chromatid   In eucaryotes one of the two identical strands of a newly replicated chromosome.

Chromosome   The molecule of nucleic acid (in procaryotes and viruses) or the complex of DNA, RNA and protein (in eucaryotes) carrying a linear array of genetic information.

Circularly permuted DNA   A population of linear DNA molecules produced as if by breaking open circular molecules at different points.

Cis-arrangement   Describes the situation where two mutant sites or genes are on the same molecule of DNA (for example, *lacZ1 lacZ2*/F' *lacZ$^+$* or *lacZ$^+$ lacY$^+$*/F' *lacZ$^-$ lacY$^-$*).

Cis–trans test   A genetic test to determine whether two mutations are in the same or in different genes.

Cloning   The production of very many genetically identical molecules of DNA, cells or organisms.

Codon   The three consecutive nucleotides in RNA or DNA encoding a particular amino acid or signalling the termination of polypeptide synthesis.

Cohesive ends   The single-stranded and base complementary sequences at the two ends of a λ chromosome.

Co-integrate (CI)   A circular molecule of DNA formed during replicative transposition by joining together two separate circular replicons.

Competence   The transient physiological state necessary before a bacterial cell can adsorb transforming DNA.

Complementation   The ability of two mutant genes to make good each other's defects when present in the same cell but on different molecules of DNA.

Concatamer   An end-to-end (tandem) array of identical DNA molecules.

Concensus sequence   An idealised nucleotide sequence where the base at each particular position is the one most frequently observed when many different actual sequences are compared.

Conditional lethal mutant   A mutant able to grow under one set (permissive) of environmental conditions but lethal under different (restrictive) conditions.

Conjugation   The establishment of a bridge between two bacterial cells and the transfer of DNA from one cell to the other.

Conjugative plasmid A bacterial plasmid able to promote conjugation.

Conservative transposition A transposition event where the transposable element is lost from its original location and reappears at a new location.

Constitutive gene expression A gene or operon which is expressed at all times and under any environmental conditions.

Copy number The number of copies of a particular plasmid present in a bacterium.

Co-transduction This occurs when two closely linked genes, or two mutant sites within the same gene, are transduced together.

Crossing-over The reciprocal exchange of genetic material to produce genetic recombinants.

Curing The treatment of a bacterium so that a resident prophage or plasmid is lost.

Deletion The loss of one or more bases or base pairs from a molecule of DNA.

Denaturation The treatment of a molecule of double-stranded DNA so that it separates into its two component single strands.

Density-gradient centrifugation A sensitive technique for separating related molecules which have slightly different densities.

Derepression The release of a gene or operon from repression so that it is expressed.

Diploid A eucaryotic cell or organism in which the chromosomes exist in pairs.

Direct repeats Two identical (or nearly identical) nucleotide sequences sometimes separated by a sequence of non-repeated DNA, for example

$$5' \quad ATCA \ldots ATCA \quad 3'$$
$$3' \quad TAGT \ldots TAGT \quad 5'$$

D-loop or displacement loop A structure formed when one strand of a duplex DNA molecule is displaced by a single strand of partially homologous invading DNA; the reaction is catalysed by the RecA protein.

DNA, deoxyribonucleic acid A macromolecule usually made up of two antiparallel polynucleotide strands held together by weak hydrogen bonds and with deoxyribose as the component sugar.

DNA gyrase A topoisomerase that removes supercoils from DNA by first producing double-strand breaks and then sealing them.

DNA ligase The enzyme that joins together the 5' and 3' ends of polynucleotide chains by the formation of a phosphodiester bond between them.

DNA polymerases The enzymes that polymerise deoxyribonucleotides on to an existing polynucleotide chain using the complementary strand of DNA as a template.

DNA primase The enzyme that normally synthesises the RNA primers required for initiating DNA synthesis.

DNase An enzyme that breaks down DNA.

Dominant   The gene (or, more correctly, the character) that is expressed in a heterozygous or partially heterozygous cell (for example, $lacZ^+$ in $lacZ^+/F'\ lacZ^-$).

Endonuclease   An enzyme that makes breaks in a molecule of DNA by hydrolysing internal phosphodiester bonds.

Episome   Formerly used to describe a genetic element that could exist either as an autonomous entity or inserted into the continuity of the bacterial chromosome (for example, F-plasmid).

Error correction   *See* **mismatch repair.**

Eucaryote   An organism with a nuclear membrane and certain organelles, such as mitochondria, and a mitotic spindle.

Excision repair   A type of repair system where nucleotides are removed from a damaged strand of DNA and replaced by a new tract of DNA synthesised using the undamaged complementary strand as a template.

Exonuclease   An enzyme that digests a molecule of nucleic acid by removing successive nucleotides from the 5' or 3' end.

F-plasmid or F-factor   The fertility factor of *E. coli*; it enables bacterial cells to conjugate.

F-pilus   The filamentous appendage encoded by the F-plasmid; it provides the adsorption sites for the male-specific phages.

F-prime, F′   An F-plasmid carrying a segment of the bacterial chromosome.

F$^+$ cell   A bacterial cell harbouring an F-plasmid.

Frameshift   A mutation which adds or deletes one or two base pairs from a coding sequence in a molecule of DNA, so that the genetic code is read out-of-phase.

Gene   The genetic unit of function; it may (i) encode a polypeptide, (ii) encode a molecule of ribosomal or transfer RNA, or (iii) be a regulatory sequence involved in the control of gene expression.

Gene conversion   An event that produces abnormal segregations by non-reciprocal recombination; it is commonly studied in ascomycetes, where it occurs following meiosis as the result of the formation of a mismatched base pair.

General(homologous) recombination   The normal system for genetic recombination; it is RecA dependent and occurs only between DNA molecules with identical or near identical nucleotide sequences.

Genome   The complete gene content of a cell or organism.

Genotype   A specific description of the genetic constitution of an organism.

Haploid   A cell or organism with only one set of chromosomes.

Heteroduplex   A molecule of double-stranded nucleic acid where the two strands are of different origin and do not have exactly complementary base sequences; such molecules are regularly produced during recombination.

Heterozygous   A cell or organism where different alleles are carried by the two members of a pair of homologous chromosomes.

Hfr strain   A strain of *E. coli* with an integrated F-plasmid, able to conjugate and to transfer its chromosome at very high frequency.

Histones   Basic proteins found complexed with the DNA in eucaryotic chromosomes.

Homoduplex   A molecule of double-stranded nucleic acid where the two strands have exactly complementary base sequences.

Homologous recombination   *See* **general recombination.**

Homozygous   A cell or organism where the same allele is carried by each member of a pair of homologous chromosomes.

Illegitimate recombination   Any type of aberrant recombination event occurring between non-homologous sequences and *not* involving site-specific recombination.

Incompatibility (of a plasmid)   The situation when two different plasmids are unable to coexist in the same cell.

Inducible system   A regulatory system where the genes are only turned on and enzymes synthesised when the appropriate substrate is present.

Induction   The switching on of transcription in an inducible system following interaction between the inducer and the repressor protein (for example, the *lac* operon).

Insertion sequence (IS)   A transposable nucleotide sequence that encodes only functions related to its own transposition.

Intercalating agent   A flat molecule that can insert between two adjacent base pairs in a molecule of double-stranded DNA, distorting the architecture of the double helix.

Intermolecular recombination (or transposition)   Recombination (or transposition) between two separate molecules of DNA.

Intramolecular recombination (or transposition)   Recombination (or transposition) between two different sites within the same molecule of DNA.

Inversion   A DNA rearrangement where a sequence of nucleotides is inverted in relation to the rest of the molecule.

Inverted repeats (IR)   Where the sequence of nucleotides along one strand of DNA is repeated in the opposite physical direction along the other strand; inverted repeats are commonly separated by a tract of non-repeated DNA. For example

$$5'\ \ GATG\ldots CATC\ \ 3'$$
$$3'\ \ CTAC\ldots GTAG\ \ 5'$$

Invertible segment   A specific segment of DNA which undergoes regular changes of orientation.

Kilobase (kb)   A unit length of 1000 nucleotides or 1000 nucleotide pairs.

Lagging strand   The strand of newly replicated DNA that is synthesised discontinuously and away from the replication fork.

Leading strand   The strand of newly replicated DNA that is synthesised continuously and towards the replication fork.

Ligation   The formation of a phosphodiester bond between two adjacent bases separated by a single-strand break.

Linkage   The tendency of genes close together on the same molecule of DNA to be inherited together.

Linkage map   A map, assembled from recombination data, showing the order of mutant sites and genes along a chromosome (or other molecule of nucleic acid).

Locus   The site on a chromosome (or other molecule of nucleic acid) where a particular gene is located.

Lysogen   A bacterial cell carrying a phage genome as a repressed prophage.

Lytic response   When an infecting phage replicates, matures and eventually lyses the bacterial cell, releasing free phage.

Male-specific phages   Small phages that only adsorb to receptor sites on the F-pilus.

Merozygote, merodiploid   A partially diploid bacterium, carrying both its own chromosome and a chromosome fragment introduced by conjugation, transformation or transduction.

Messenger RNA (mRNA)   The transcript of a segment of chromosomal DNA which is a template for polypeptide synthesis.

Micrometre (μm) 1×10^{-6} metres, 1000 nm.

Mismatch repair, error correction   A mechanism that corrects mismatched base pairs that have escaped correction by the proofreading activities of the DNA polymerases.

Multiplicity of infection (MOI)   The ratio of phage to bacteria in an infection.

Mutagen   An agent that increases the frequency of mutation.

Mutant   A cell or organism with a defective gene and displaying a specific altered phenotype.

Mutation   The process that produces a sudden heritable change in the nucleotide sequence of an organism; any such change in the nucleotide sequence.

Mutator gene   A mutant gene which increases the frequency of mutation in other genes.

Nanometre (nm) 1×10^{-9} metres.

N-glycosylases   Repair enzymes that are able to excise certain incorrectly paired or damaged nucleotides.

Nonsense mutation   A mutation which alters a codon for an amino acid to a codon for chain termination.

Nucleoside   A purine or pyrimidine base linked to a molecule of ribose or deoxyribose.

Nucleotide   A nucleoside with an attached phosphate group.

Okazaki fragment   The short discontinuously synthesised fragments that are ligated together to form the lagging strand during DNA replication.

Open reading frame (*orf*)   A sequence of in-phase codons preceded by a translational initiation codon and terminated by a chain termination triplet.

Operator   The DNA sequence to which a repressor protein reversibly binds so as to regulate the activity of one or more closely linked structural genes.

Operon   A group of coordinately controlled structural genes, together with the operator and promoter sequences that control them.

Origin (*ori*)   The nucleotide sequence at which DNA replication is initiated.

Palindrome   A pair of adjacent inverted repeat sequences.

Permissive conditions   The particular environmental conditions which allow the growth of an organism with a conditional lethal mutation.

Phage (bacteriophage)   A virus that infects a bacterial cell.

Phage induction   A treatment which stimulates prophage to enter the lytic cycle; eventually the host cell lyses and releases free phage.

Phenotype   The appearance or other observable characteristics of an organism.

Phosphodiester bond   The covalent bond joining the 3'-hydroxyl of the sugar moiety of one (deoxy) ribonucleotide to the 5'-hydroxyl of the adjacent sugar.

Photoreactivation (PR)   The repair of ultraviolet-irradiated DNA by the cleaving apart of pyrimidine dimers.

Plasmid   An extrachromosomal, covalently closed, circular molecule of DNA carrying non-essential genetic information and replicating independently of the bacterial chromosome.

Point mutation   A mutation involving the addition, deletion or substitution of a single base pair (or sometimes several adjacent base pairs).

Post-meiotic segregation (PMS)   The type of segregation produced when a recombinant DNA molecule contains an uncorrected mismatched base pair; at the next replication, normal base pairing occurs producing one mutant and one wild type product.

Post-replication repair (PRR)   A repair process, probably involving a recombination event, which enables the survival of a DNA molecule that is damaged at the same site on both strands.

Primase   The same as DNA primase.

Primer   The short strand of RNA which provides the free 3'-OH end on to which DNA polymerases can then add deoxyribonucleotides.

Procaryote   An organism lacking a nuclear membrane and certain organelles such as mitochondria.

Promoter   The DNA sequence to which RNA polymerase binds in order to initiate transcription.

Prophage   A phage genome whose lytic functions are repressed and which replicates in harmony with the bacterial chromosome; prophage is frequently (as with λ) integrated into the chromosome of the host cell.

Prototroph   A micro-organism able to grow on minimal medium containing only a carbon source and certain inorganic salts.

Pyrimidine dimers   Ultraviolet irradiation of DNA has caused covalent bonds to form between two adjacent pyrimidines on the same strand.

r-determinant   The component of an R-plasmid carrying the gene(s) for drug resistance.

R-plasmid   A transmissible plasmid encoding resistance to one or more drugs.

Reading frame   One of the three ways by which a coding sequence of nucleotides can be read in consecutive groups of three.

Recessive   The gene (or, more correctly, the character) that is not expressed in a heterozygous or partially heterozygous cell.

Recombinase   An enzyme promoting genetic recombination.

Recombination   *See* **General recombination, Illegitimate recombination** *and* **Site-specific recombination.**

Replication fork   The Y-shaped region of a DNA molecule where the two strands have separated and replication is taking place.

Replicative transposition   The transposition of a transposable element to a new location without its being lost from the original location.

Replicon   A DNA molecule able to initiate its own replication.

Repressor   The protein product of a regulator gene; when it binds to its own operator it prevents transcription of the associated structural genes.

Resolvase   An enzyme which resolves a co-integrate molecule into its two component replicons.

Restricted transduction   A system of transduction where only genes closely linked to an integrated prophage can be transduced.

Restriction endonucleases   A large group of endonucleases each of which recognises and attacks a specific sequence in a DNA molecule; they are extensively used in recombinant DNA technology.

Restrictive conditions   The particular environmental conditions which do not allow the growth of a conditional lethal mutant but result in the expression of the mutant phenotype.

Reversion   Any mutation that restores the wild phenotype of a mutant.

Ribosomes   Complex assemblies of RNA and protein, the sites of protein synthesis.

RNA polymerase   The enzyme that normally synthesises RNA against a DNA template.

RNase   An enzyme that hydrolyses RNA molecules.

Rolling-circle (σ) replication   A type of replication where a replication fork moves round and round a circular molecule of DNA producing a single-stranded concatamer. This single-stranded product may become double-stranded by the synthesis of a complementary strand.

RTF, resistance transfer factor   The component of an R-plasmid encoding the ability to conjugate and to transfer DNA.

Same-sense mutation   A mutation which changes the nucleotide sequence of a codon so that it still encodes the same amino acid.

Selection   A method, using a special set of environmental conditions, which only allows the survival of mutant or recombinant cells with a particular phenotype.

Sex-pilus *See* **F-pilus.**

Silent mutation A mutation which changes the nucleotide sequence but does not cause a detectable change in the phenotype.

Single-strand binding protein (ssb) The protein that binds to the single-stranded DNA at a replication fork, protecting it from nuclease attack and from reforming double-stranded DNA.

Site The position of a mutation within a gene; a specific base pair.

Site-specific recombination (SSR) A special type of recombination occurring only between short specific DNA sequences and requiring special enzyme systems.

SOS box The operator sequence recognised by the LexA repressor protein.

SOS response A set of functions, including an error-prone repair system, induced by the presence of damaged DNA.

Suppressor mutation A mutation that restores, partially or completely, the loss of function caused by another mutation. Many suppressor mutations are in genes encoding a transfer RNA species; the altered tRNA can recognise the original mutant codon and, during translation, insert an acceptable substitute amino acid into the polypeptide.

Tandem duplication A directly repeated DNA sequence.

Temperate phage A phage that can either establish itself as prophage or enter the lytic response when it infects a sensitive bacterial cell.

Tetrad The four products of a single meiosis.

Tetrad analysis A method for establishing linkage relationships by analysing the four products from individual meiotic divisions.

Theta-(Cairns-) type replication Replication of a circular molecule of double-stranded DNA by initiation at a unique origin and proceeding in one or both directions around the molecule.

Trans-arrangement Describes the situation where two mutant sites or genes are on different molecules of DNA (for example, $lacZ1/$F′ $lacZ2$ or $lacZ^+ \; lacY^- /$F′ $lacZ^- \; lacY^+$).

Transcription The process by which RNA polymerase (or DNA primase) produces a molecule of RNA using one strand of a DNA duplex as a template.

Trans-dimer synthesis An error-prone process which permits nucleotides to be inserted opposite a pyrimidine dimer without their being excised by the proofreading activity of the DNA polymerases.

Transductant A genetic recombinant formed during transduction.

Transduction The transfer of genes from one bacterium to another by a phage vector.

Transfection Transformation using purified intact phage DNA.

Transfer RNA, tRNA Adaptor molecules which, on the one hand, bind to a particular amino acid and, on the other hand, recognise the corresponding amino acid codon on the messenger RNA, thus correctly aligning the amino acids for assembly into a polypeptide.

Transformation   The transfer of genes from one bacterium to another by 'infecting' a recipient strain with purified DNA extracted from a genetically different donor.

Transition   A base substitution mutation where a purine replaces a purine, and a pyrimidine replaces pyrimidine.

Translation   The assembly of amino acids into polypeptides using the genetic information encoded in the molecules of messenger RNA.

Transposase   A specific enzyme required for the transposition of a particular transposable element.

Transposition   The transfer of a discrete segment of DNA from one location in the genome to another.

Transposon   A transposable genetic element which, in addition to encoding the proteins required for its own transposition, confers one or more new recognisable phenotypes (usually resistance to one or more specific drugs) on the host cell.

Transversion   A base substitution mutation where a pyrimidine replaces a purine and vice versa.

Vector   An independent replicon, usually a small plasmid or viral genome, used to introduce a 'foreign' gene into a host cell.

Virulent phage   A phage that can only enter the lytic cycle when it infects a sensitive bacterial cell.

FURTHER READING

General

Bainbridge, B. W. *Genetics of Microbes,* 2nd edn, Blackie, Glasgow (1986).

Brown, T. A., *Gene Cloning, an Introduction,* Van Nostrand Reinhold, Wokingham (1986).

Campbell, A., *Episomes,* Harper and Row, New York (1969).

Glass, R. E., *Gene Function* — E. coli *and its heritable elements,* Croom Helm, London (1982).

Hayes, W., *The Genetics of Bacteria and their Viruses,* 2nd edn, Blackwell, Oxford (1968).

Lewin, B., *Gene Expression, Vol. 1 – Bacterial Genomes,* Wiley, New York (1974).

Lewin, B., *Gene Expression, Vol. 3—Plasmids and Phages,* Wiley, New York (1977).

Lewin, B., *Genes,* 3rd edn, Wiley, New York (1987).

Miller, J. H. and Reznikoff, W. S. (eds), *The Operon,* 2nd edn, Cold Spring Harbor Laboratory, New York (1980).

Neidhardt, F. C., Inghaham, J. L., Low, K. B., Magasanik, B., Schachter, M. and Umbarger, E. (eds), *Escherichia coli and Salmonella typhimurium — Cellular and Molecular Biology,* American Society for Microbiology, Washington, D.C. (1987).

Schleif, R. F., *Genetics and Molecular Biology,* Addison-Wesley, Reading, Massachusetts (1986).

Stent, G. S. and Calendar, P., *Molecular Genetics, an Introductory Narrative,* 2nd edn, Freeman, San Francisco (1978).

Watson, J. D., Hopkins, N. H., Roberts, J. W., Steitz, J. A. and Weiner, A. M., *Molecular Biology of the Gene,* 4th edn, Benjamin/Cummings, Menlo Park, California (1987).

Zubay, G., *Genetics,* Benjamin/Cummings, Menlo Park, California (1987).

Collections of papers and articles

The following collections of reprinted papers include many that have made significant contributions to the development of microbial and molecular genetics.

Abou-Sabé, H. A. (ed.), *Microbial Genetics*, Dowden, Hutchinson and Ross, Stroudberg, Pennsylvania (1973).

Adelberg, E. A. (ed.), *Papers on Bacterial Genetics*, Little Brown, Boston and Toronto (1960).

Drake, J. W. and Koch, R. E. (eds), *Mutagenesis*, Dowden, Hutchinson and Ross, Stroudberg, Pennsylvania (1976).

Gunsalus, I. C. and Stanier, R. Y. (eds), *The Bacteria, Vol. V, Heredity*, Academic Press, New York (1964).

Haynes, R. H. and Hanawalt, P. C. (eds), *The Molecular Basis of Life — Readings from the* **Scientific American**, Freeman, San Francisco (1969).

Stent, G. S. (ed.), *Papers on Bacterial Viruses*, Little Brown, Boston and Toronto (1960).

Taylor, J. H. (ed.), *Selected Papers on Molecular Genetics*, Academic Press, New York (1965).

Tomizawa, J-i (ed.), *Bacterial Genetics and Temperate Phage, Vol. 1*, University Park Press, Baltimore (1971).

Zubay, G. L. (ed.), *Papers in Biochemical Genetics*, Holt Rinehart and Winston, New York (1968).

INDEX

Page references in *italics* refer to Figures, Tables and Boxes.